PRAISE FOR

Rescue and Revival

"Through visionary strategic planning, deep relationships with talented and generous people, and a real sense of drama, Gregory Long transformed NYBG over his thirty-year tenure into a vibrant world-class institution, having an absolutely wonderful time along the way. This is an inspirational story of a man's total commitment to one of the country's most important and beloved cultural institutions. A treat to read."

—Betsy Smith, President and CEO, Central Park Conservancy

"Gregory Long reveals what it takes to manage and safeguard a key public institution . . . how to be a maestro . . . and a fascinating insight into leadership."

—Thomas E. Lovejoy, Senior Fellow, The United Nations Foundation

"Each of Gregory Long's many achievements as president of NYBG is impressive on its own, but brought together in *Rescue and Revival*, they make for a remarkable tale of taking a venerable but shaky institution leaps and bounds toward its full potential of greatness. Through these captivating personal stories he teaches us how a cultural institution can become more valuable to both its local and international communities, inviting visitors, students, and scholars to learn plant biology, save biodiversity, and find out how plant research can mitigate climate change."

—Gloria Coruzzi, Carroll and Milton Petrie Professor, NYU
Member, National Academy of Sciences

Rescue and Revival

LALH

Rescue and Revival

NEW YORK BOTANICAL GARDEN

1989–2018

Gregory Long

FOREWORD BY

Robin Lane Fox

LIBRARY OF AMERICAN LANDSCAPE HISTORY

AMHERST, MASSACHUSETTS

Library of American Landscape History
P.O. Box 1323
Amherst, MA 01004
www.lalh.org

Printed in the United States of America by Porter Print Group, LLC

Library of Congress Control Number: 2021947912
ISBN: 978-1-952620-37-9

Designed by Jonathan D. Lippincott
Set in Galliard

Distributed by
National Book Network
nbnbooks.com

Frontispiece: Rock Garden Cascade in spring.
Photo © and courtesy Larry Lederman.

All photographs courtesy New York Botanical Garden unless otherwise noted.

Rescue and Revival was made possible with the generous support of:

Leon Levy Foundation
Anonymous
Susan P. Burke
Larry Condon – New York Community Trust
John William Bernstein and Diana Davenport – Vanguard Charitable

Contents

Foreword

by Robin Fox Lane

Rescue and Revival is an unmissable account of a great success story. The title is too modest. More than a revival, Gregory Long's transformation of the New York Botanical Garden made it the admiration of public garden stewards all over the world. Adventurous, exciting, but always sympathetic, the process was far more than a well-planned makeover. Long's book records innovation, commitment, and again and again, apt pride and love by everyone involved.

Anyone with major responsibilities for a public garden will gain from reading this book. So will the ever-increasing ranks of fundraisers worldwide and, I hope, donors and would-be donors, on whom great cultural institutions depend. It explains what is possible, how to do it, and why. Naturally, *Rescue and Revival* will also appeal to the fans of NYBG, a massively enhanced constituency which is still growing. It will add depth to what they enjoy on visits to the garden by setting this important place in the wider perspective of history, from its origins as ancient woodland and rocky outcrops to its transformation into the world's most watched botanical garden in an age of ever-shrinking public funding. For me, visits to New York City are a sort of purgatory until I can escape to the green living heaven in the Bronx, NYBG and all it offers to redeem me. It is a pleasure, therefore, to write this foreword.

My own experience mirrors the book's story in miniature. I first visited the garden in January 1980 when it was not exactly at its best. In downtown Manhattan my hosts advised me to make the short road-crossing from the rail station in the Bronx to the garden's main gate as rapidly as I could, for my own safety. But I made the crossing helped, not harassed, by Bronx residents who were only too willing to assist a stranger with spectacles not always strong enough to see signs at a distance.

Inside the garden, I admired the magnificent allée of tulip trees that ran up as a long, leafless avenue to what I now know to be a world-famous library, graced with treasures made possible by the likes of J. P Morgan and Andrew Carnegie. In the crisp chill of a January afternoon, I took shelter in the glasshouse, though several panes of glass in the roof were cracked and had been blocked over with temporary cardboard.

In April 1997, I tried again, happy to find there was now a charming tram system to take me to the further reaches of a site so much bigger than I realized. On an idyllic spring day, I remember swathes of narcissi and, in the Rock Garden, an excellent specimen of the lilac-flowered *Daphne genkwa,* which I had never managed to grow in England. Times had certainly changed. In September 1998, I returned again, this time to explore the conservatory in its restored splendor: it made my old memories seem quaint. I wrote in the *Financial Times* about the excellence of the newly displayed collections and the audio guide that had led me to plants of significance without being bossy.

This visit was to be crucial for my relationship with the garden, in ways I did not foresee. Several weeks after my article appeared in the *Financial Times,* I received a letter of thanks, handwritten, from Gregory Long, appreciating what I had written about the Enid A. Haupt Conservatory, the massive project whose planning, funding, and execution are so vividly described in this book. I was struck by the trouble given to this letter, and a year or so later was equally struck when I met the author at a lunch he hosted in London for English friends of the garden. Readers of *Rescue and Revival* will not need to be told of the impression his modesty, humor, manners, taste, and mastery of his responsibilities made on me and all present.

For more than fifteen years I have been fortunate to have enjoyed the friendship and personal example of Gregory Long. One thread in this book is Gregory in less-expected contexts, his early work for the New York Zoological Society being one. "Animals had never," he writes, "been my highest personal priority." We follow him through the Metropolitan Museum of Art and on into the New York Public Library, where his friendships with the legendary Brooke Astor and his great role model, Vartan Gregorian, are memorably set out. He would not claim to have been a major gardener, let alone a botanist, when he moved on to NYBG in 1989. Yet it was to be a match made in heaven—primarily, of course, of his own making. "We were aiming at world-class standards," he remarks.

When NYBG began, its cofounder Nathaniel Britton cast Kew Gardens as his model. By the 1990s many botanical gardens, including Kew, were looking on gardening as secondary to their priorities of science and research. At the same time, they were losing all or most of their public funding. In New York, Gregory and his board had been quicker to realize that they must function at the highest level in all three areas: science, gardens, and education. By 2010, NYBG had become the envied role model for public engagement, fundraising, and gardening, closely studied by Kew itself.

I have moved on the edges of fundraising and donations in a lifetime spent in an Oxford college and its university and have often witnessed the random outcomes of kind engagements by donors. Out of the blue they provide a new monument or bit of stonework. An unexpected professorship in an unexpected field is established. One, even two, institutes of climate change or aging appear. But there is a disjunction between what the powerhouse—the teachers and thinkers—and the bureaucracy want to do. Led by Gregory, NYBG's approach has been so different.

As he describes in the book and its appendix, he prepared priorities for each campaign by beginning at the bottom of the organization and working to the top before producing a plan over several years. It is not one imposed from the top, like those Soviet five-year horrors, but one agreed to by the people at every level who had to work with it. Gifts then were directed to the needs established in advance, which were then wholeheartedly carried out.

What gifts they have been. Not just the amazing gift from Enid A. Haupt to ensure the rescue of the conservatory or LuEsther Mertz's for the library, but gifts for the excellent visitor center, which reoriented access to the garden; the superb American native plant garden, dotted with glorious scarlet lobelias in July; the brilliant replanting of the azalea area; the underplanting of the woodland; the collections of lilacs and of maples; and the completion of the triangular rose garden designed by the great Beatrix Farrand but never fully realized until donors, curators, and Gregory's presidency made that possible.

Nine hundred million dollars of private sector donations and another $300 million from public sources tell their own stunning story, to the credit of the givers, requesters, planners, and putters into practice. Gregory also takes us into less-visible sides of the enterprise, its crucial scientific mission transformed by major gifts so as to engage with the possibilities opened by molecular studies. As a gardener and writer, I have followed two other sides in its progress with delight: the educational and the horticultural.

"Charity begins at home" many of us used to be taught in Britain. Quite often the result was that charity never reached beyond the front door. As Gregory well explains, he was heir to a tradition of civic philanthropy, given to fine institutions in one or another city. I have heard outsiders, ignorantly, allege that NYBG is a social hub for New York's hyper-rich with no real interest in gardens except as venues in which to drink champagne. What nonsense, as this book excellently shows. I single out NYBG's Green-Up project and its growing and sharing of food with local residents in the Bronx, a project which continued during the 2020 lockdowns when other public gardens around the world had no such network. The latest link in that chain, the remarkable Edible Academy, is a brilliant step, bringing at least 100,000 people a year into serious contact with vegetable gardening for the first time. Those whom I have known on the NYBG board have real commitment and delight written all over them, not least through the projects which carry the garden and its skills far out into society.

Civic philanthropy is in ever more competition with global giving and the hope to alleviate problems in a faraway world. Global giving appeals to young, mobile givers, but civic giving must not lose out, because it can be effectively managed and targeted to enduring local effect. NYBG has thought laterally, enlarging its impact. Since 2010, I have been able with Gregory and the board's kind help to visit and write freely on its groundbreaking summer exhibitions. I had long marveled at the blindness of many fine art experts, museum curators, and literary scholars to the place of flowers and plants in poems, works of art, and artists' own lives. The NYBG summer shows, as Gregory has envisaged them, addressed that gap in cultural understanding and rebuilt the link between flowers and gardens and artists and individuals.

In the conservatory the long walkway became an artistic stage for Francisca Coelho, suprema of the wonderful displays she somehow sustained through the hot months of summer. Books and paintings, loaned for the occasion, linked the library into the event, while poetry boards, to my delight, brought the flowers and gardens in the show into relation with what its main subjects had written. The Emily Dickinson exhibit in 2010 became a fine duet between her poetry and love of flowers. The Frida Kahlo show in 2015 presented the painter's little-known courtyard garden in Mexico together with a series of loaned pictures by her, beautifully displayed in the library gallery. The daring Monet show in 2012 spilled out triumphantly into the conservatory's courtyard and lily pool, under Coelho's brilliant styling. These exhibitions have not just been a big new source of yearly income. They have brought hundreds of thousands of new visitors into the garden and made them aware, whether in poetry, art, dance, or cuisine, of plants' and gardens' links with surrounding culture. We live more alertly as a result.

Through these shows I have come to regard as friends some of the vice presidents whom Gregory has fostered, whether Francisca Coelho and her unrivaled cool under glass, Todd Forrest and his vast range of practical knowledge and planting outdoors, or Melinda Manning, so discreet in her help with press and publicity. There are many others, and it is so true when Gregory refers to them in this book as a family.

He sketches a Perfect Visitor to NYBG (setting a standard of which I fall dreadfully short). I would like to end with a Perfect President, one whom its board, vice presidents, and staff will recognize. He begins by having the president's excessive apartment on Park Avenue sold for the garden's benefit. He draws up detailed strategic plans but knows the name and background of everyone who works under his aegis, whether sweeping the paths or caring for the magnificent herbarium he has helped to preserve. He looks outward, beyond America, as those of us know who have been guests at the Garden's fine dinners for the Chelsea Flower Show, and as the Garden's style attests. He is a bit of a producer at heart, loving to stage a party where people enjoy meeting each other, often through his own initiatives. He has a firm grip on future directions and current spending. He attracts trustees and donors because he really likes them, and they really like him. They want to join in.

At the president's last Conservatory Ball, Gregory was honored by celebrities, supporters, board, staff, and even the mayor's office, which declared Gregory Long Day. On the morning of the ball he had taken time to take me around the garden. So much must have been in his own mind, yet he talked only of the importance of institutions and of the history and the landscape which made this garden special. It was the best tribute to what he never said, all that it owed to him personally in thirty stunning years.

Rescue and Revival

PROLOGUE

A Morning in Copenhagen

One chilly November morning in Copenhagen, when visiting that wonderful city for the first time, I walked from my hotel to the Botanical Garden to have a look. It's a lovely late nineteenth-century place, surrounded by the low-rise city, and used by neighbors for walking and sunning on such a morning. At the edge of the parklike landscape stretches a sprawling white wood conservatory that houses tropical and desert plants. The building was closed, but I felt a kinship with this place with its urban setting and Victorian aesthetic—so like ours in New York. It was a great age for botanical gardens.

After exploring quietly for an hour or two, admiring the finely maintained grounds and the old conifers, I stopped near the entrance at the little gift and book shop—a stop I always make in sister institutions. That morning candles were burning in hurricane lamps on the steps of the shop, as they are at the entrances to many Copenhagen establishments. Charming! Alluring! As always I was drawn to the bookshelf. There I discovered one rather academic-looking volume from the 1920s, *Botanisk Have Gennem 50 Aar, 1874–1924.* The author was C. H. Ostenfeld, director emeritus of the garden, who had written an account of his years there. Although it was quite dated, and almost exclusively a summary of scientific explorations and work, I thought to myself, "Aha! As I approach the end of my long tenure at NYBG, I think a similar project might be a useful thing to undertake." This book is the result of that moment, not an official or authorized history, but my account of twenty-nine years in the life of the New York Botanical Garden, from my first day as president and CEO until my retirement in 2018.

INTRODUCTION

Change Was the Order of the Day

When I arrived, the New York Botanical Garden was a distinguished but sleepy, even neglected, organization, like a historic house or garden with good bones but falling apart, in need of a major renovation. It was a large institution—even in those days it took 250 employees to run—and although it had been named a National Historic Landmark, it was in bad shape. The old physical plant had had almost no maintenance since the early 1900s, the 250-acre landscape was simply sad, and the reputation of the place as a cultural institution in New York was in tatters.

The trustees who hired me knew they needed and wanted change, and they asked me to engineer it. Over nearly three decades I assembled and led the team—trustees, volunteers, professional staff—that did the work to breathe new life and meaning into an organization that in some important respects had lost its way and was misunderstood by most of the people of New York City. For me and my colleagues in this enterprise, it was the most rewarding professional experience of our lives despite its feeling at times like a hard slog, a frustrating and sometimes disappointing obstacle course.

But we did it! We persevered, repairing and rebuilding, renovating and rethinking, from top to bottom, including programmatic objectives and activities and finances as well as the physical plant. We restored the spirit and dignity of the place. We expanded and energized our powerful board of trustees and enlarged our staff to five hundred full-time employees. Through this process, NYBG became a prestigious scientific and horticultural institution commanding international respect. From an organization with no fiscal strength, it became a financial powerhouse that could claim a twenty-nine-year history of balanced budgets of close to $90 million, an endowment of more than $350 million, and unrestricted net assets of $500 million.

Over the course of those years we raised $1,200,000,000, invested much of it, and spent the rest wisely. We built or restored dozens of programmatic facilities to better serve our burgeoning audience, which grew by sixfold. We restored and developed more than half of our 250 acres of historic landscape and gardens, and we reduced our carbon footprint. Everyone in New York and elsewhere saw that the Garden had become truly beautiful. And throughout it all, we never wavered from the noble mission of our founders: to study the plants of the world and to share our knowledge with the public.

We all feel now that we have made a huge difference—to the Bronx, to New York City as a whole, to the causes of scientific research and science education for children, and to the American appreciation of good horticulture and nature. It was thrilling to take an institution that was in many ways moribund and fix it! And it is this excitement and sense of accomplishment that I have tried to convey in these pages. I think my readers will be people who care about their museums, cultural institutions, maybe even their social service and religious organizations, and perhaps they will be inspired to take on tasks similar to ours and make a difference themselves—volunteers, philanthropists, and professionals of all stripes. My hope is that they will benefit not only from the lessons embedded in these chapters but also from the enthusiasm for a seemingly impossible task accomplished against all odds.

It's a different world today from what it was thirty years ago, but the essential tasks we undertook to resuscitate NYBG are like the tasks facing many not-for-profit stewards in the post-pandemic period—or any period in the future. At the Garden, we were not only repairing an organization that had fallen on hard times, we were also imagining a new and more exciting future for our enterprise, and we were expanding its role and moving outside our comfort zone to reach new and different audiences. It was at once rescue and revolution.

If you are interested in how a cultural institution can grow and change, become more financially secure, and make itself better and more valuable to its communities—in our case, Bronx neighbors, urban children with no outdoor experience, New York City culture-goers, students and scholars wanting to learn plant biology and save biodiversity, scientists needing to find out how plants can mitigate climate change, and people abroad who live in the places where their livelihoods are endangered because of environmental degradation—I think you will enjoy my stories.

I have included a great deal of New York City social history in this account because many people involved with NYBG in my era were significant figures in the life of New York during these decades. Whenever possible I recruited important people in our field of interest (for example, E. O. Wilson, the eminent American biologist, was with us all the way), in the civic life of the city (Mrs. Vincent Astor was a great friend to the institution), and in philanthropy (Lewis B. Cullman, the father of our molecular and genomics programs, joined with us at a time in his life when he had determined to give all of his great wealth away). These people and many others walk in and out of these pages, in ways that I hope will be both amusing and illuminating to the reader. They were all major players in the process.

ONE

A New York Cultural Life, 1969–1989

Over these past years I have delivered many public lectures and taught classes on the New York Botanical Garden: the historic landscape, our programmatic interests, or various aspects of our management practice. At the end of these sessions, a question is often asked, "Could you share with us the reasons why you were chosen to lead NYBG—what was your path to this wonderful place?" I'm always a little bit taken aback, because the question implies that perhaps I was not qualified, not being a professional botanist or horticulturist. But I have grown used to answering it.

I started out on my road to the presidency of the New York Botanical Garden in September 1969. From that day until I began work at NYBG, I had the challenge and excitement of working in the high echelons of administration at five other major cultural institutions in New York City: twenty years of "graduate education" and practicum. While I was waiting on that September day for my interview in the luxurious outer office of Ashton Hawkins, Esq., corporate secretary of the Metropolitan Museum of Art, I may have had a vague sense that this might be the beginning of a career that would end up at the helm of one or another New York cultural institution. But mainly I was just a really young guy needing a job and hoping to work at the Met. I had graduated from college that spring, and except for painting houses, mowing lawns, and shelving books in an art reference library, I had never had a job. I had just finished my undergraduate work at New York University with a bachelor's degree in the history of art. NYU and its graduate Institute of Fine Arts was home to many German classical art historians who had emigrated to New York before, during, and after World War II, and I had come to study with that distinguished faculty, focusing on Renaissance art and architecture and the ancient world.

As I waited for my interview, I had in mind that I would work for two or three years to gain some experience and then return to the Institute for a graduate degree. I had thought that I would like to be a scholar of Italian Renaissance art. But the truth was that the IFA's German language requirement was too tough for me. And I was beginning to feel that the life of a scholar might be less active and less relevant than I wanted. I had had a traditional upbringing in the Midwest, and my intellectual interests and emotional connection were to history. But it was 1969, and I was among the youthful protestors demonstrating against police violence in Chicago and the war in Vietnam. Although I was not by any measure a student radical, I did understand that the sixties had happened and the world was in many ways transformed. I knew we were living in a modern world, and I was eager to make a difference.

The interview went well, and rather miraculously, just weeks before my twenty-third birthday, I was offered a position as executive assistant to the corporate secretary of the Metropolitan Museum of Art, at a then-handsome salary of $11,000. I stayed at the Met for about two and a half years, working hard in that corporate office and learning a tremendous amount about museums and institutional management, and the art of engineering change at an established place. The experience has served me well throughout my entire professional life.

Much of what we came to do at NYBG was based in concepts and skills I learned at the Met. The museum was at that time, early on in the administration of Thomas J. Hoving, a well-oiled machine, and it was on the brink of a new era. George Trescher, a brilliant agent of change, was in charge of the centennial celebration, a yearlong event staged in 1970 with tremendous fanfare, intended to draw international attention. The Met was in the throes of transformation. It was being democratized—an old and elite aura of scholarship and wealth was shed, and the doors thrown open to the people of New York City and the world. New programs, new departments of contemporary art and ethnographic material were established, and blockbuster exhibitions of American art, medieval art, costume history, and mid-twentieth-century New York School painting were mounted on a scale never seen before.

The administrative and curatorial staffs were serious and highly energized; they worked together in synchrony and with mutual respect. The board, under the leadership of C. Douglas Dillon, former secretary of the Treasury in the Kennedy administration, was powerful and impressive. In my position as executive assistant, I was expected to write business letters and other correspondence for the signatures of Mr. Hawkins and Mr. Dillon, and I sometimes wrote the first drafts of their presentations and speeches. I'm not sure exactly how I learned how to do this, but I copied letters in the files and I did careful research with colleagues in the curatorial ranks and in the administration. I was a small player, but I was able to observe closely how a really well run museum works. I ate it up.

I learned to be an editor from Mr. Dillon, and I learned to deal with people of great wealth and power without fearing them. As it was part of my job, I taught myself to forge Douglas Dillon's signature on his letters and memos. Readers may remember that, for many years in the 1960s, U.S. currency was signed by C. Douglas Dillon, Secretary of the Treasury—I could do that signature! Another of my responsibilities was the organization and preparation of board meeting materials, and through this work, I learned not-for-profit corporate governance backward and forward. These materials were voluminous documents in those days, all hand-typed, and I was in effect the captain of the typing pool. I learned to do this by copying materials I found in the files and by quizzing the women in the typing pool, one of whom had been doing it for years. Those women, by the way, could type 100-plus words per minute perfectly, and in those days, on an old-fashioned IBM Selectric typewriter, I could do 75—just not perfectly. I also observed creative exhibition planning from both marketing and curatorial points of view, and I was exposed to the new concept in not-for-profit life of corporate sponsorship.

I made many friends at the Met with whom I maintained lifelong relationships. One was Henry Geldzahler, the much-admired curator of twentieth-century art who had recently established that department; another was Dietrich von Bothmer, curator of the Met's significant collection of Greek and Roman art. They each in their own way taught me about museum life. During this time, I also casually befriended Mrs. Vincent (Brooke) Astor. Mrs. Astor was one of the most important museum trustees, and as I was dealing with the trustees before, during, and after their meetings, we would sometimes encounter each other around board meetings. Aside from saying, "Good morning, Mrs. Astor," I wouldn't speak to her unless I was spoken to, but she was very pretty and flirtatious and I think I was rather flirtatious as well, so we began, when I was in my early twenties and she her late sixties, a relationship highly insignificant then but years later very significant to me.

After more than two years of sixty-hour weeks, I decided—for reasons not entirely clear to me to this day—to resign from this wonderful job and look around for something else to do. I think I was worn out. And I really did not like dealing with my super-demanding boss, Ashton Hawkins, or with Tom Hoving, an arrogant and foul-mouthed man. Anyway, I quit.

It wasn't entirely bad judgment. Almost immediately I was offered an extremely interesting eight-month assignment at the Brooklyn Museum. During my years at the Met, I had become friends with two senior curators at Brooklyn, and they asked me to come "operate" their new special exhibitions program. Following the lead of the Met and its successful centennial exhibitions, the Brooklyn Museum was trying its hand at (little) blockbusters. They mounted a Norman Rockwell retrospective (which I never liked but was popular) and a beautiful exhibition of their fine collection of American prints and drawings. The museum had the in-house expertise to create and curate these exhibitions but no one on staff who could actually run them on a daily basis. Back then there was

Mrs. Vincent (Brooke) Astor, c. 1975.
Photo by Francesco Scavullo.

no such thing known nowadays as visitor services—I was to become Visitor Services. This seemed to me a fantastic opportunity to learn about running a museum from the front of the house as opposed to the executive suite. So I bought a new blue blazer and drove my old Volkswagen every morning, six days a week, from my apartment on the Upper West Side diagonally downtown, across the Brooklyn Bridge, and over to the museum.

I would arrive every day fifteen minutes before we were to open and meet the head of security in the main lobby. We would switch on all the lights, telephone the accounting office for the cash box to be brought up to the volunteer behind the card sales desk, and, at 9:59, unlock the front door. We would throw the doors open, and I would begin a long day of pacing through both exhibitions to make sure everything was going smoothly. We would make sure that all the bulbs in the lighting system were working, that the racks of brochures were stuffed with reading material, and that the pictures were hanging straight. I would check that the volunteer behind the desk was keeping things tidy (and not smoking a cigarette). There would usually be one security guard on duty for most of the day, but I was the floor walker. As in a department store. I loved meeting and talking with visitors. Since I was wearing an official museum badge, I was directing them to the restrooms, encouraging them to buy postcards, sending them around the bend to the museum shop, which in those days was a great place to buy antiques, and generally filling them in on the life and work of Norman Rockwell in informal gallery talks. So I was also a docent.

It was an invaluable experience in what I am pleased to call my education because, unlike my work at the Met, which had been about governance, finance, and high-end thinking, this was about the people who used the museum. It taught me that the light bulbs had to be in place, the floors had to be swept, the bathrooms had to be clean, and the doors had to be opened at ten o'clock sharp—it seemed to me a performance akin to a Broadway play. The curtain went up every morning at ten, and I was conducting the orchestra. This is a concept I have never forgotten—it is an aspect of museum management that I love. And over my years at NYBG, I tried to instill this idea—of the curtain going up, everyone onstage in their place, the Garden in perfect condition—to all of our employees responsible for opening the gates each morning.

The Brooklyn job ended in early 1972, and after a little looking around and consulting with George Trescher, who had become a mentor, I landed at the American Museum of Natural History. Trescher was an independent consultant at this point in his career, and he was advising there on the creation of a new fundraising program. I accepted the position of director of development, about which, truth be told, I didn't know very much. I had not worked in fundraising at the Met or the Brooklyn Museum, but here I found myself in a job that was 100 percent fundraising. But it actually didn't matter that I didn't know anything, because neither did anyone else at the museum. There had never been a fundraising office there, just two elegant gentlemen, the museum's president (the head of the board and a volunteer) and vice president (a former magazine publisher who was responsible for the museum's business activities). They hired me, and as a result I was what in the navy would be called "staff" not "line." I reported to the president of the board (not too different from my position at the Met). I did not answer to the director.

With the help of several board members, especially Howard L. Clark, the longtime chair of American Express (whose signature I also learned to forge because I drafted and signed his letters—readers old enough to have had American Express traveler's checks in the 1970s may remember that they were signed by Howard L. Clark), we ramped up some old-fashioned committees that had done very polite fundraising through tea parties, added new members to those groups, and improved their solicitation materials and methodologies. In addition, we created new programs to bring potential donors into the museum to learn about its fascinating scientific research activities and engaged the renowned anthropologist Margaret Mead as a spokesperson. We expanded the board and launched an annual corporate-giving program, with a large business committee and an elaborate letter-writing campaign in which industry leaders would solicit funds from their peers. The museum was much-loved, and inventing this new fundraising program was not actually very difficult. I would say it all fell into place. There had been a large membership program, and, of course, that extensive list of modest donors who cared about the museum's programs was a great foundation, fabulous grist for our mill.

I was enjoying this assignment and learning an enormous amount about fundraising strategies and materials, when one day in 1975, the director of the museum, to whom I did not report, asked me to come to his office (it was my first time ever in that room), and he fired me. He said something insulting and vaguely sinister like, "We really don't want people of your sort here in our museum." Since my little staff and I were raising millions of dollars, all new dollars, for his museum, and I was having a very happy time with my bosses the president and vice president, you can imagine I was taken aback. I went downstairs to the president's office and complained. But he was an elderly gentleman, a good fundraiser but not a strong person, and he told me that he was very sorry, I had done an excellent job at the museum but that he could do nothing for me. Insulted, disappointed, I left, although not entirely unhappily.

I loved the museum's complex juxtaposition of scientific research institution on the one hand and its role in public exhibition and education on the other. (Later, when I arrived at NYBG, I found an analogous situation—a botanical research staff behind the scenes totally unappreciated by the general public.) I had become a good spokesperson for the importance of basic research in the biological sciences. But I had never felt quite at home there.

In only a matter of days, however, in the summer of 1975, I was offered an exciting position at the New York Zoological Society, the international conservation organization that operated the Bronx Zoo (and is now known as the Wildlife Conservation Society). Once again, George Trescher had recommended me for the job, building from scratch new membership, fundraising, education, and audience development programs for the Bronx Zoo and the New York Aquarium. All of a sudden, I was a vice president, reporting to the director of the Society, William G. Conway. Conway was rather formidable, the top zoo man in the world, but he was creative and he was bringing the Bronx Zoo and their international programs out of the dark ages. This was another institution in the throes of dramatic change, and I understood that they wanted me to bring my experience in institutional development and transformation, much of it learned at the Met, to support their new identity. Conway was taking the animals out of their cages and putting them in naturalistic habitat exhibitions (and he was a gifted designer of these exhibitions himself—especially the beautiful ones of birds). He wanted to make it clear to the world that zoos are meant to inspire visitors to become activists in wildlife conservation.

It was a fascinating effort for this old and antiquated place, and although I had never felt myself to be a "zoo person," I was eager to accept the challenge of creating this new programming because, of course, I understood that zoos are museums without walls. I built a large team of new hires, all young and smart and fun to work with, and Conway and the trustees supported my every request for the funding to support these new enterprises. I was part of the team that cleaned up the zoo, changed the visitor demographics to include city and suburban families looking for entertaining but educational outings, and greatly expanded that new audience. We put on marvelous weekend special events to generate publicity and new visitors—Easter egg rolls, the "Animals' Thanksgiving," safari tours through "Wild Asia." And once a year we staged a highly amusing evening for members at Avery Fisher Hall, replete with live animals (elephants, even, and boa constrictors) and films of wildlife conservation projects abroad. It was an important annual event in New York City—thousands attended.

Fundraising is always laborious, but (as had been the case at AMNH), finding new funding for the Society was not too difficult because there had been no fundraising at all before—it had simply been old-line trustees passing the hat. And early on at NYZS, I made friends—for real—with Brooke Astor. Mrs. Astor was on the board of the Zoo. She loved animals, and she adored being a trustee. Her best friend on the board was Laurance S. Rockefeller, brother of Vice President Nelson Rockefeller and David Rockefeller, the famous head of Chase Bank. Laurance Rockefeller was an avid advocate for wildlife conservation, and as soon as I got my fundraising program up and running, he and Mrs. Astor became my best partners. There was no dinner they wouldn't chair, no friend they wouldn't try to recruit to NYZS membership, no one they wouldn't ask for money. They both really liked to work at building up the organization, I guided them, egged them on, and supported them, and we had terrific success and great fun.

Toward the end of my seventh year at the Society, I began to feel a little restless—animals had never been my highest personal priority. It was 1981, and the new president of the New York Public Library, Vartan Gregorian, a fabulously charismatic man who had been recruited from the University of Pennsylvania, was much in the news. Somehow I got word to him and his chief of staff that I would be interested in coming to work for him to create the same kinds of new programs—fundraising, membership, education, exhibition, and audience development—inside NYPL that we had launched at the Zoo. Luckily, just at that time Gregorian had fired the vice president for development he had inherited, and he was looking for new blood. I was asked to meet with him one afternoon at three o'clock, and we went, curiously enough, to the empty upstairs dining room at Sardi's Restaurant in the theater district. It was dark in that room, but it was a scintillating meeting—he captivated me with his highly original and exciting ideas about moving NYPL into the center of the city's cultural, literary, and social life. It was 1981. He offered me a senior vice presidency, and of course I jumped at it. I loved books and research libraries, and I was enticed by the idea of building and expanding New York's constellation of neighborhood branch libraries (which number eighty-eight today).

But there was a hitch. Brooke Astor and I were having so much fun and success together at the Zoo, and she loved the Zoo so passionately, that I was afraid she might be unhappy about the idea of my leaving. As it happened, she was a trustee—just about the most important one—of the New York Public Library too, and there she was again. I knew I had to ask for her support. So, one fine day I asked if I could come visit her at her sumptuous apartment on Park Avenue. We had tea, and I explained to her, with considerable trepidation, that I was interested in a new and exciting position at the New York Public Library. I was afraid she wouldn't like it. But she

Vartan Gregorian, c. 2005.
Photo courtesy Carnegie Corporation of New York.

jumped to her feet, beaming, and practically shouted, "Oh, darling, do you want me to get it for you?" What a relief! No, I said, thank you, I can get it for myself, I just came here today to ask you not to oppose it. She laughed and assured me that not only would she not oppose it, she would support it wholeheartedly. I laughed and thanked her, and I think we moved on right away from tea to scotch and soda. Nice afternoon.

Gregorian and I made an incredible team. He was the front man, and no one was ever better at that role. With his flamboyant charm and intellect, he became the darling of New York media and society, all to the benefit of NYPL. With me behind the scenes, we worked together for seven years and accomplished an enormous amount. We restored—inside and out—the Carrère & Hastings Central Research Library at 42nd Street; invented a successful large-scale exhibition program based on the Library's research collections and important literary subjects; created a program of public lectures and symposia which became a model for many other organizations; and raised hundreds of millions of dollars to improve conditions at the Schomburg Center for Research in Black Culture, the Library for the Performing Arts at Lincoln Center, and the branch libraries throughout the city. The New York Public Library shook off its dusty past and became a celebrated and vibrant place.

Mrs. Astor, affectionately known by almost everyone in New York as Brooke in the 1980s and '90s, was a truly catalytic figure, the undisputed leader of New York civic, philanthropic, and social life. Born in 1902, she lived 105 years. She was a woman of tremendous intelligence and charm, known for her superb judgment about people and city affairs, pretty well-loved by almost everyone. Governors, senators, mayors, international statesmen, captains of industry, bankers, writers, art historians, TV personalities, socialites, people like me working in cultural institutions were all ardent admirers of Brooke Astor, and we were all at her dinner parties. She was a widow when I first met her in the 1970s. She had had three husbands—the first, a man she divorced; the second, a man she loved deeply and who died in her arms; and third, the enigmatic Vincent Astor, great-great-grandson of John Jacob Astor, the first multimillionaire in America, to whom she was married during the 1950s. When he died in 1959, his very substantial fortune was divided between a large bequest to her, much of which she was required to distribute to charity at the time of her death, and a large private foundation. Mrs. Astor was chair of the Vincent Astor Foundation for more than forty years—and she gave it all away. In addition to being a creative and thoughtful philanthropist and a society hostess, Mrs. Astor was also an author. She loved books and wrote four of her own—two novels and two memoirs. She was a human being of very great talent in so many different ways. She was also, as you may have guessed, extremely glamorous—practically a movie star.

Vartan Gregorian and Brooke Astor were way, way out front, and I was in the office, planning and plotting, writing a new future for this incredibly important institution. It was a great honor to be in this role at NYPL and to work with them for so many years—and it was also barrels of fun. One of my many responsibilities was to partner with the top finance man to develop a program of strategic planning for the entire library system. We invented a bottom-up process that began with librarians, curators, and administrators working on the front lines and filtering up through middle management to upper management and then to the trustees. We developed programmatic plans, capital plans, and financial plans, all interdependent. It was exhilarating. Even long-term and rather jaded library employees were excited, as were the trustees, as were our friends at City Hall, as was the philanthropic community. When we finished the fundraising campaign that followed the strategic planning, we had raised $300 million, pretty much evenly spread between public and private sources—the largest amount of funding ever generated for an American library system. It was the go-go eighties, of course, and New York was ready for what Gregorian and NYPL offered. One marvelously creative employee of ours, when asked what she would be doing in her new assignment at the Library, answered that she thought she would get in the Library and drive it up Fifth Avenue. Which we did.

Much to my regret, in 1988, Gregorian accepted the presidency of Brown University and left NYPL. I was bereft—it had been a once-in-a-lifetime partnership. We had invented so much together—a new brand and position in the community for the Library, exhibition and public education programs that were the envy of New York institutions and libraries everywhere, a huge strategic planning program copied by many other institutions, and a large-scale fundraising program—and I wasn't ready to quit. I wanted him to stay forever and ever, and I wanted to stay there too. An international search for his replacement was begun, directed by Andrew Heiskell, chair of the board, and without encouragement from anyone, I decided to compete for the position. I was forty-two and I thought it was a long shot, but I loved the Library and really didn't want to leave, and I thought I could probably do it. However, I never made it to the short list, and Andrew, who had become a great friend (and would remain one for another twenty years), took me to lunch one day at a diner downtown and explained that I didn't have the stature he wanted in Gregorian's successor. He said he hoped I would continue as vice president, but I think I told him that would be rather unlikely.

We are now in 1988, and just at this time the New York Botanical Garden was involved in a search to replace its retiring president, James Hester. Once again, Brooke Astor played a role. This is one of the only New York cultural boards she was never on (unfortunately, as she would have loved it), but of course there were many board members who were her friends and friends of her closest associate, Linda Boyer Gillies, the longtime CEO of the Astor Foundation. I don't believe I ever asked Linda to recommend me, but she did, and I later learned that after she had spoken to NYBG trustee Beth Straus her recommendation had gotten me a meeting with the search committee. I enjoyed my several meetings with that committee—all men, despite the fact that two women were at that point probably the most important trustees, Beth Straus and Enid Annenberg Haupt.

Beth Straus, whom I knew a little, was a brilliant idea woman who had recruited many important people to join the NYBG board in the seventies and eighties. In my era, she became senior vice chair of the board and was hugely creative and helpful. Her best friends on the board were Thomas J. Hubbard, Mrs. David (Peggy) Rocke-

Enid Annenberg Haupt, c. 1985.

feller, and Mrs. Haupt. The daughter of a wealthy publishing family, the Annenbergs of Philadelphia, Enid Haupt was a widow, the most important patron ever of American horticulture, and far and away the most important benefactor ever to the New York Botanical Garden. In the 1970s she had given the major private money required to restore the conservatory, in recognition of which the conservatory had been named in her honor.

But Mrs. Haupt was not an easy woman. There were many people on the board who found her impossible to deal with, and as a consequence, this most important figure was not a member of the presidential search committee. In December 1989, after many, many meetings and exploratory conversations, I was offered the presidency. I was overwhelmed and thrilled (although I was brokenhearted in some ways to think about leaving the Library), and I wanted to accept, of course. However, I had known Mrs. Haupt a bit over the years, especially during my time at the Library, and I was concerned that she had not been one of the decision makers offering me the position. I decided that I would go see her and ask her opinion. I thought she would be flattered, and I hoped that would go some distance toward our establishing a productive working relationship, but of course I wasn't sure. I was sure, however, that if for some reason she were opposed to my taking the job, I would have a very hard time of it.

It was December, and she was in residence in Palm Beach, where she always spent the cold months. I phoned and asked if I could come visit for an afternoon to talk about the future of the Garden. She agreed, and I flew down, knocked on the door of her mighty mansion, and was let in by Pedro, her tiny and ever-loyal butler. As I sat waiting in the courtyard, I thought, this is a risky business. But soon she came in, beautifully dressed and coiffed, and all smiles. She seemed to know I had been offered the job, and when I asked her what she thought, she said I should accept. I never knew whether she *really* supported me or whether she was just flattered she was being included in the decision making. Not that it really mattered.

Afterward we went for a walk in her extensive garden, and when I gushed that I found it fabulous, she replied, "Well, what did you expect?" Some months later, after I had accepted the position and was preparing to start, at an informal meeting of some trustees, she exclaimed for all to hear (and believe me, when she spoke, the trustees of NYBG listened), "I believe that Mr. Long will be a great success in this position, because he knows, and I know, that if he isn't—his career is ruined." The truth is, that from that day in Palm Beach until the day her executors called to tell us that she had bequeathed to NYBG her entire wealth, $108 million, and in spite of our ups and downs over many years, I had always known she was with us, with me.

TWO

Starting Out: Dogs and Cars

And so it was that after two decades of assignments in New York museums and a research library, I found myself at my second museum without walls, this time at the helm. From the start I reveled in learning the history and position of NYBG. I came to understand that this was a museum of plants. Also through my reading, I learned that at times in the Garden's twentieth-century history, this essential fact had been ignored or perhaps misunderstood by the leadership of the institution. Since my background is in museum management, and I am a museum person at heart, I was excited to reestablish this concept right away.

I learned that over the past centuries, really since the Renaissance in Europe, botanical gardens have been created by and for people who care about plants. First there were professors of medicine searching for new cures. Then there were explorers bringing home to Europe exotic species to delight the curious and to supply patrician plant collectors. There followed the scientists determined to survey, understand, and classify the natural world, and in the nineteenth century, builders of empires wanting to "borrow" economically valuable plants from indigenous people for imperial commercial benefit. Of course, throughout time, botanical gardens have also provided sustenance for passionate gardeners who love to grow whatever they can and enjoy what they cannot.

By the time NYBG was founded in emulation of the Royal Botanic Gardens at Kew, botanical gardens were home to both academic scientists attempting to understand biodiversity and educators and social reformers interested in exposing city dwellers to the wonders and beneficial qualities of nature. Established to serve the international scientific community in its quest for new knowledge, NYBG was also created to teach the public to value and love plants—to be an institution of both research and public education. I was immensely proud to be responsible for its future.

I came to understand that over a century of effort, scientific research at NYBG had developed into a massive scholarly enterprise. And, outside, in the Garden itself, our programming was mainly addressed to plant and garden lovers. Over my twenty-nine years at NYBG, as we continued to think about our educational role, we expanded our vision to create public exhibitions and programs relating horticulture to culture and science to art. We decided to be of value to people who didn't necessarily care about plants or gardening at all. We looked beyond our core audience to a broader group whom we could convince to care about plants by showing them Charles Darwin's charming garden at his home in the English countryside, or Frida Kahlo's jazzy courtyard garden in Mexico City. Or by introducing them to garden design at the Alhambra, to medicinal plants in the early round botanical garden in Padua, Italy, or to the plants of Hawai'i in the paintings of Georgia O'Keeffe. In our major special exhibitions, we combined plants and gardening with the lives and work of cultural figures, and we explored historic gardens. We exhibited cultural artifacts and works of art in our galleries, and in our great conservatory we mounted floral spectacles reproducing in living color gardens from around the world. As a result, we discovered a vast new audience of individuals who before hadn't known they loved plants. Without giving up any of our research interests or activities, we expanded the meaning and identity of the botanical garden as a public museum. But in 1989, I did not know we would be moving in this important new direction. It was to come later.

What I did know was that I was now the leader of an organization established by an act of the New York State legislature in 1891.

Nathanial Lord Britton; Elizabeth Knight Britton in the first-growth forest at NYBG.

The creation of a botanical garden in the City of New York had been a dream of the scientific community and parks advocates since at least the mid-1880s. (See John Mullaly, *The New Parks beyond the Harlem*, 1887.) The Garden was the brainchild of Nathaniel Lord Britton, a natural history professor at Columbia University, and his wife, Elizabeth Knight Britton, an eminent botanist. Married in 1885, the couple took their honeymoon in London in 1888. Because they were both interested in plants (Elizabeth Knight Britton was an expert on mosses, and N. L. Britton, on the trees of North America and the plants of the Caribbean), they visited the famous Royal Botanic Gardens at Kew, then the world's greatest botanical garden. Legend has it that they were overcome by the beauty of that historic landscape, the impressive glasshouse for tropical plants, and by the significance of the institution. As members of the powerful New York Torrey Botanical Club, named for the distinguished nineteenth-century botanist John Torrey (whose papers later came to be housed in the library of NYBG), they were among those who understood that a botanical garden would be a considerable asset to New York City and also a boon to them in their professions.

Excited and more ambitious even than I was a century later, the young couple returned to New York and began a public campaign for the creation of the new garden. Elizabeth Knight Britton was especially active in orchestrating the movement. Along the way she enlisted many of New York's richest and most influential men in the cause. The founding trustees of the New York Botanical Garden were all important figures in New York business, finance, law, and science. They included Cornelius Vanderbilt II, John D. Rockefeller, Andrew Carnegie, J. P. Morgan, and Judge Addison Brown—all leaders in city and state politics and life. Following the creation of Central Park in the 1850s, this was the period when the city's ruling class was determined to make New York a significant and civilized metropolis, comparable to London, and established many of the city's great cultural and educational institutions, including the Met and the American Museum of Natural History.

The site chosen for the Garden in 1891 was a 250-acre parcel in the northern section of land that had been acquired in 1888–89 for preservation as a park. The picturesque landscape resembled the

The Garden landscape under construction, c. 1895.

beloved Hudson River Valley, with the Bronx River flowing through, elephantine rock outcrops, and first-growth forest. This location was selected partly for the venerable fifty-acre forest at its heart, and, as much of the land had not been farmed, it was also the site of hundreds of mature deciduous and coniferous trees—some of which are alive today—and other native plants. These precious aspects of the original landscape formed the first, foundational, "collections" of the living museum. But, of course, museum collections are not created by God or nature—as I knew very well, they are created by scholars and curators. At NYBG, these second-level collections are the taxonomic or scientific, first laid out across the 250 acres in the 1890s and expanded every generation since.

The curated collection of NYBG represents about forty thousand taxa, or types of plants, one of the largest in any of the world's botanical gardens. This is a cosmopolitan, comprehensive, circumpolar collection—temperate, tropical, and desert plants. We try to be encyclopedic. The collections are arranged sometimes taxonomically (for example, the Lilac Collection and the Magnolia Collection) and sometimes aesthetically (the Perennial Garden, the Home Gardening Center), but more often in a blend of intentions both taxonomic and aesthetic (the Azalea Garden, the Rose Garden, the Million Daffodils). Over the decades the most important groups of plants assembled by the horticultural experts working at NYBG have included the desert plants of Africa and North America, palms, cycads, tropical orchids, tropical blueberries, conifers grown from seed collected from nature, the native plants of the northeastern United States, and shrubs and trees including azaleas, maples, oaks, magnolias, cherries, and crabapples. These so-called taxonomic collections have been assembled and curated by professional staff for study by plant experts, designers, and the lay public enjoying the variety of nature.

But not all of our collections are "living," and early on I came to understand and respect the herbarium. These pressed and dried plant specimens have been assembled over many generations by scientists working in the field, and they are the core collection supporting the institution's basic research programs in plant science.

Angraecum sesquipedale ("Darwin's orchid"), one of thousands of rare orchids in the NYBG collection. Photo © and courtesy Larry Lederman.

There are now 7,900,000 specimens, millions of which have been digitized and are available to students and scholars worldwide.

The collections of the institutions where I have spent my professional life have all been different, but they all have one central function in common: to collect and to exhibit and interpret material for the purposes of research and education at all levels. In the most fundamental sense, museums are places where collections are arranged for the purpose of comparative study, even if nowadays (as is the case at NYBG) new audiences are coming for a cultural experience only tangentially related to the museum's collections.

When I arrived at NYBG, the trustees and I knew there were potentially dangerous situations to be addressed. It is often the experience of a new museum (or botanical garden) director that some emergencies have to be dealt with up front. As I began to quench these fires, I surveyed my available staff to see who could help. Very fortunately indeed, the NYBG president's office was managed by two exceptionally talented women, Mary McLaughlin and Celeste

Kirton. They had been my predecessor's assistants, and they were a gift to me. Smart, charming, seriously witty, experienced at NYBG (Mary for two decades already), fearfully hard-working, and on top of all that, they both took dictation: Gregg shorthand! Mary stayed for twenty-five of my twenty-nine years ("It's time," she told me), and Celeste and I both had our last day at NYBG on June 30, 2018. I knew my good luck and never took them for granted—we had decades of the most satisfying work and lots of laughs.

I saw that I needed a new security director, and one always needs a new fundraising staff. But there were other gems in addition to Mary and Celeste already on board, especially a very determined young man, John Rorer, who had come to the Garden only a few months before my arrival, as vice president for finance and operations. He was an experienced nonprofit financial administrator and an excellent professional. We liked each other—we needed each other! John had been at the Garden long enough that he could brief me immediately on our precarious financial situation. During the search process I had researched the Garden's position in many respects, and I had already begun to analyze the financial picture. I knew it was grim. But at least I had a partner.

On my first official day, John was in my office before 10:00 a.m. The Garden's fiscal year was about to close on June 30, and John had identified a gap between income and expenses of approximately one million dollars. The fundraising staff could not see any way to find this money, especially not in eight weeks. I'm sure I shared this difficult fact with key board members, but none of them had any thoughts on how to solve the problem.

So, what did we come up with? When he came to work at the Garden in 1981, my predecessor, James L. Hester, PhD, had been given the use of a large, NYBG-owned apartment on Park Avenue, in the lower 90s, as his official residence. He and his wife Jan had lived there for his entire nine-year tenure. When I had accepted the presidency, I was sent to have a drink with the Hesters to see the apartment. At this time, I was living in my own very comfortable place on West End Avenue and 85th Street—a nice drive (in those days) up the West Side Highway to the Garden. I had never been interested in living on the Upper East Side. And it also didn't seem to me that this starving institution needed to provide its CEO with such a luxurious benefit. So, John and I decided that I would stay ensconced on West End Avenue and we would sell the Park Avenue apartment.

John discovered (maybe this was my second piece of good luck), that the apartment had been bought ten years earlier with funds taken from the operating budget, which meant that if we could sell it, the funds could be returned to the operating budget. The apartment was worth slightly more than a million dollars. John leapt to action, and we were able to sell that apartment in the last few days of June—balancing the budget with the proceeds. Imagine how relieved we were. I had been in office just sixty days and because of John's ingenuity and fast work, we had balanced the budget! This was the first of twenty-nine consecutive balanced budgets from 1989 to 2018. It was also the first effort by our new administration to put the house in order.

A second problem we encountered at the start, which I was aware of from my research during the search process, was the balkanization of the staff. The horticulture or gardening staff was jealous of the scientific research staff—they felt like second-class citizens. The scientific research staff, composed principally of academic PhD scientists, was proud of their distinguished history and work, and they felt they deserved most of the (limited) money. The poor education staff was just left behind, which often happens in museums that are not well run. The operations and security departments were desperately underfunded and treated like lower-class citizens. The administration? There wasn't much. The conflicts among these factions were principally over money—there just wasn't enough, and there was a lot of anxiety and grousing about who would get what.

This situation was to some degree reflected among the board: the men mainly were supportive of the scientific research function, and the women mainly supported the horticulture side. Some board members (of both sexes) understood the robust tripartite nature of the institution and were levelheaded; some were not. I believed that I could begin to solve this problem by a fair-minded approach to management, but I also knew that solving such a serious institutional situation would require a deeply thoughtful process of strategic planning, and I began in those first few months to reassure both staff and board that it was a problem we could solve.

Another area John Rorer and I had to attack in the first few months was the management of Garden security. The primitive stockade fence surrounding the property was in very bad condition, filled with gaps which neighborhood teenagers made use of at night. The beautiful rock garden had been fenced with razor wire to keep the marauders out. I learned that our security staff had seized guns and other weapons from Garden visitors over the years, and that there was a collection of these weapons in the security office. At one of the regularly scheduled all-staff meetings we established, unbeknownst to me, some of the security men displayed these guns, knives, and blackjacks on a table in the lobby outside the lecture hall—they were actually proud of their cache. John and I were appalled, and we realized we needed a new head of security, and completely new thinking about how to keep the Garden safe.

So we got rid of all those weapons and tried to instill in the security staff the idea that the Garden was to be known as a safe zone. The incumbent head of security retired to Florida, and we hired Robert Heinisch, an ambitious and smart young man who had been recommended to us by the head of security at the Bronx Zoo. It took Bob about a month to solve many of the problems we had been faced with. Never again would there be a gun inside the Garden. The fence was repaired, the razor wire removed, and a canine unit established for nighttime patrol. The entrance gate was locked in the evening and guarded by trained personnel.

Once again, I had been lucky—Bob was exactly the right man for the job. He was also a gifted choreographer of Garden visitors. Tall and handsome, well mannered, and insistent, he could stand in the middle of the lawn on a crowded Sunday afternoon and show people how to behave in a museum of plants. He eliminated ball playing, an activity that belongs in parks. He restricted picnicking to the picnic areas. He prevented groups of teenagers from running amuck. He and I discussed Garden etiquette at length, and we decided that we would simply require that Garden visitors behave in a respectful fashion.

So, with Mary and Celeste and John and Bob in place, I had my first fine team. Our next challenges were about the buildings. There had been generations of deferred maintenance and questionable judgment with regard to the function of some facilities. For example, there were two apartments inside the Garden, one in the historic Lorillard Stable and one in the 1840s Stone Mill, a National Historic Landmark. These apartments were old and filthy, and they were, of course, illegal, and our insurance didn't cover us for liability or damage. It wasn't easy to deal with the inhabitants—an obstreperous family with a dog in the Stone Mill and a much-respected senior scientist in the Stable—but somehow, we did it. When it came to the upstairs apartment in the Stone Mill, we were faced with the most massive house cleaning imaginable: mains, drains, bathrooms, plus faulty electrical service. We did a gut rehab of the entire second floor, above what we called the ballroom, a beautiful space used for special events and fundraising (which we restored in 2010). Also in the Stone Mill there was a problematic restaurant in the basement. Kitchen conditions were a nightmare. Grease-choked drains, and river rats on the terrace where the staff and Garden visitors had lunch! The restaurant did serve exceptional gravlax, but we knew we had to eliminate the operation, and although it took about seven years, we finally did.

One of the things that I and other people involved with the Garden found especially rewarding was that our task was in so many ways one of restoration. The board and many staff were dismayed that in the 1970s and '80s, NYBG had lost its stature among the city's cultural institutions. Scientific research, the library, and the herbarium were keeping up to international standards, but the Garden's profile in New York City had diminished greatly. The landscape was degraded, the living collections were being expanded only modestly, and the Garden's premises were abused by its users. During its long period of neglect, this museum of plants had been transformed into a large public park—dog walking, automobiles cruising through,

Tulip Tree Allée. Photo © Marlon Co.

families playing frisbee—and it was fairly lawless. In late 1989 we invented a program we called "Spring of '90." We wanted to throw out the dogs and the cars and create a Garden-wide program to restore the dignity of the place.

The roads winding through the Garden had been modernized in the early 1920s by the Olmsted Brothers landscape architecture firm. Suitable for the traffic in those days, over the succeeding decades, these narrow roadways had become thoroughfares carrying big American cars, and too many of them. Visitors were driving down Azalea Way, across Daffodil Hill, and around through the magnolias, sometimes parking along the verge to reorganize the stuff in their trunks. Neighbors would bring their cars over and clean them along the side of the roads. And the road climbing the hill to the library, which was defined by a magnificent double allée of American tulip trees, was lined with parked cars. I admit that when I first started I parked there myself.

Then there was the ball playing, the frisbee tossing, and, worst of all, the dog walking. There's more acreage of parkland in the Bronx (because of its development in the later nineteenth century) than in any other borough of New York City, but the Garden had become a favorite place for dog walkers. During meetings of the presidential search committee, there had been much discussion about these unhappy conditions and inappropriate uses. It was very clear that something had to be done. Although we had very little money, we decided we could, in a few steps—and before too many people noticed its decline—turn the New York Botanical Garden back into a museum of plants. We could restore its pride. Some wonderful friends volunteered to help, among them the landscape architect Patrick Chassé, the distinguished New York architect Hugh Hardy, and the public garden designer Lynden B. Miller.

We started at the perimeter—graffiti on the stockade fence. Patrick Chassé taught us that we should paint the fence and then paint over the graffiti every morning when it had reappeared. We chose a beautiful dark green, a "gardenesque" color, and we started spraying. We removed parking from Tulip Tree Allée. We created a new Garden map to help walkers navigate through the complicated landscape and spent the winter months of 1989–90 constructing tall pylons with the map and interpretation, designed by Hugh Hardy.

We also made another superb hire in these early days, a freelance contractor and carpenter, Mark Cupkovic, who had done some work at my West End apartment. We learned that he was looking for an institutional position that would give him more security than his freelance life provided, and John Rorer hired him as director of operations. Neither of us knew how brilliant Mark would turn out to be, but I believe he was the most talented person in such a position at any cultural institution in New York City. Mark and I shared an interest in good housekeeping, and we began repairing the Garden's buildings and infrastructure. Both of us were "house-proud" all through our time together—he retired about a year after I did. It was a great partnership.

There was horticultural housekeeping to be done across the Garden's entire 250 acres as well, and I plunged deep into horticultural management. Applying everything I knew about gardening and tree care, everything I had learned from my own garden and from years of garden visiting in Europe, I worked with our horticulture team to raise the bar considerably in the maintenance of the Garden's landscape. I became very good at spotting weeds under herbaceous perennials and dead branches high in specimen trees. We were aiming for world-class standards. In addition, we bought, and brought in staff to operate, two Garden trams, each of which seated sixty people, so there would be motorized transportation throughout most of the 250 acres. No more driving over Daffodil Hill.

To achieve all of this we did have to spend some money. We raised about $200,000—which in those days was a lot for us. Generous members of the board who were eager to see these improvements were the primary donors. And we worked to change the mindset of Garden staff, volunteers, and neighbors. There was some consternation among the science staff, and we had to deal with that. Having grown accustomed to working at an institution that had lost its dignity, at first they resisted my efforts to restore beauty and a museum identity. The idea that the beauty and dignity of the whole place would actually support their work was new to them. But we met together, and—I believe largely because of the foresight and goodwill of Patricia Holmgren, then acting vice president of botanical science—they agreed to suspend disbelief, and some even became advocates of the project. They understood that despite their excellent work, the institution's international reputation was in danger.

Our most dramatic move in "Spring of '90" was to throw out the dogs and the cars. With the help of Marie Sexton, our remarkable vice president of external relations, who had come with me from NYPL, we set May 1, 1990, as the launch date for the "new institution." As of that day, we announced, when the Garden opened in the morning, there would be no cars driven beyond the perimeter parking lots, there would be no ball playing or active sports, and there would be no dogs. To be perfectly honest, I was nervous, afraid it wouldn't work. But Marie Sexton and Bob Heinisch were not to be daunted. They were true believers, they were young and attractive, they had charm, and they both had loud voices.

Starting at nine-thirty that morning, before the Garden opened at ten, they stood outside Mosholu Gate (the only entrance to the Garden at that time), and faced off with angry dog walkers. They explained that dogs damage botanical gardens and that the vast parklands of the Bronx were available to them. Gentlemen who drove up in big cars to give their mothers a ride through the Garden were told that they could park and take the tram. I think no one was more amazed than I was—after about a week of protest, the whole thing died down. We had begun the transition from abused public park to respected botanical garden.

Our "Spring of '90" program was a great success. The staff saw that they could begin to be proud of the Garden. The board saw that we could address serious problems with dispatch. I started to like my job. We had begun to clean things up, but there was so much to be done—programmatically, physically, financially. There was so much to be changed and improved, and of course, there was no future planning. It was clear to me that we needed an extensive and thoughtful strategic plan.

As the process we had invented at the New York Public Library had been the catalyst for change in that institution, I decided we should begin a similar planning exercise at NYBG. At NYPL we had planned not from the top down as is often done but from the bottom up. There had been much to do, and we attacked it in a holistic fashion. The plan took two years, perhaps more, to create, and when it was finished we went out and raised the funds—hundreds of millions of dollars from the city, state, and federal governments and the private sector—to make our plans reality. With this huge infusion of money came powerful new energy, and the stodgy old NYPL was given new life. It regained its position as one of the central cultural, educational, literary, and scholarly institutions in the country. Our strategic plan, our blueprint, was the document that made it happen.

When, in 1989–90, a few key trustees and I realized that NYBG needed the same treatment—to be "turned around"—it was logical to create a strategic plan modeled on the successful New York Public Library experience. We began by explaining to a board planning committee (composed of several trustees and the senior staff) the methodology we would use to achieve consensus on our collective future. As at NYPL, planning would be structured to include staff in all divisions and at many levels. We would plan for programmatic development over a seven-year period, decide on capital projects, and draft a seven-year financial plan that would quantify needs. We decided on a seven-year plan because there was obviously so much to be done that it didn't seem to me we could do it in five. And a ten-year plan sounded too long and too Soviet. The seven-year decision proved viable for us, although the capital work dictated by this and our future seven-year plans would, we learned, often take eight or nine years to accomplish.

Jeanne Liedtka, professor of business administration at the Darden School of the University of Virginia, calls our methodology "structuring an institution-wide dialogue." She has written three case studies based on our planning, but they do not describe "how to" do it. They were written for use in business school curricula and are not too helpful to museum administrators in terms of structuring such a process, but they are inspirational because we did in every case turn our plans into reality. The methodology we used has never been well documented, and as an appendix to this book, I have spelled out the steps we took in the creation of these plans.

Our first plan was published in 1992 as a three-part document—a programmatic description, a text on capital needs, and a financial plan—integrated and interlocking. One of the many benefits of this exercise is that it dissolves balkanization. People in the organization understand and appreciate the work of other divisions and departments. Also the CEO and financial leadership will have distributed the costs associated with programmatic improvements and capital projects in a balanced way across the organization, so it feels fair. For example, at the Garden we set essentially even caps on the seven-year costs of improvements and forward movement in scientific research and those in horticulture. Education would lag behind financially, but it was a much smaller division with many fewer employees.

A massive 1993–99 fundraising campaign was based on the plan. Four large-scale capital projects together costing about $125 million sat at the center of the effort: complete restoration of the Enid A. Haupt Conservatory; building of a large-scale food service facility adjacent to it; construction of the Steere Herbarium and ren-

The completely restored Enid A. Haupt Conservatory. Photo © Robert Benson Photography.

ovation of the interior of the historic library building to provide new space for the Mertz Library; and creation of a children's garden. So, what were we able to fund and accomplish? The story of how we generated the money for the new herbarium and library facilities is told in chapter 4, and I describe the adventure of restoring the conservatory—a project that took every minute of four years—in chapter 5.

We timed the food-service building so that it would open simultaneously with the restored conservatory in May 1997. A handsome traditional building, designed by the New York architects Cooper Robertson, it houses a large elegant ballroom for catered events with a seating capacity of three hundred, a reception room for meetings and cocktail parties, and a restaurant where lunch is served to Garden visitors. Opening the facility enabled us to close the old decrepit café in the basement of the Stone Mill. The project was funded with a $6 million gift from the LuEsther T. Mertz Charitable Trust. Several trustees felt it was a risky venture—that we shouldn't be in the food-service business. They believed that we would be better off investing the $6 million in a new endowment fund. Fortunately, we took the gamble, and for twenty-four years now our successful cater-

The Garden Terrace Room. Photo © André Maier.

ing operation has generated approximately $1.25–1.5 million in net proceeds annually, and along the way everyone saw that investing in this facility had been much more valuable than creating a new endowment. The food-service and catering programs have fed hundreds of thousands of Garden visitors every year, and brought thousands of new people into the Garden for weddings, bar mitzvahs, anniversaries, and corporate events. (Long ago, we decided not to permit other not-for-profit organizations to hold their fundraising parties in our facilities, but to reserve them for our own purposes.)

Our fourth big capital project was the Everett Children's Adventure Garden, which opened in 1998 to national and international acclaim. Situated on four hilly, rocky acres between the entrance to the historic forest and the conservatory (and near what would become, as our master planning developed, the new visitor center and main entrance to NYBG), the garden was designed to help teachers and parents teach plant science and ecology to children in an engaging, hands-on way out of doors. In this natural setting, we designed five large-scale gathering places or outdoor classrooms, each with a particular theme set to correspond with the National Science Standards for grades K–6 and the New York State science curriculum. All of these subjects had been taught at NYBG before—by schoolteachers and by our instructors, both volunteer and paid—but there had never been a dedicated outdoor facility designed to help adults introduce these concepts to young children. Each area incorporated

Entrance to the Everett Children's Adventure Garden.

Brooke Astor and Gregory Long at May 1997 opening of Everett Children's Adventure Garden.

signage and interactive exhibitry teaching the basic lessons of plant biology—"Plants Are Alive"; "What Is Photosynthesis?"; "Plant and Animal Interaction." In the center of the five-acre site were two charming buildings designed in a storybook Adirondack style (by the architectural firm of Richard Dattner) housing three classrooms, a teacher prep and resource center, and staff offices. In a way we created a botanical garden within the botanical garden. Almost thirty years later, two generations of New York City schoolchildren have come here to enjoy the outdoors and learn about subjects such as how plants evolve to adapt to their environments.

Audience development also became a principal area of focus during this period. We invented the *Holiday Train Show*, an entrancing exhibition with G scale model trains and a collection of 140 miniature New York City landmark buildings created by a team of artists entirely out of plants—bark, cork, lacquered leaves, nuts, seeds, twigs, berries. Now famous worldwide, the show has been staged every year since it opened in 1992 and brings as many as 285,000 people to the Garden in some years.

In scientific research, this was the period of the invention of our new molecular systematics program, fully described in chapter 4,

Holiday Train Show. **Photo © and courtesy Larry Lederman.**

and the restoration of the historic Arthur and Janet Ross Lecture Hall by Hugh Hardy—a 400-seat hall perfect for lectures, films, and music, designed in the late nineteenth century when the library was built, and used today as many as ten times per week for large audiences, scientific, horticultural, and civic.

As we came to the sixth and seventh years of the plan, the Garden was balancing its annual budget and our staff had grown from 261 to 387. It was a time of tremendous growth, evolution, and great excitement for the institution—it launched our era. The budget for realizing our goals had been fearfully approved by the board at $175 million, and in the end, we had raised $208 million. We had worked for seven years to generate that funding and spend it wisely, and we were ecstatic with what we had accomplished. Planning had worked, fundraising had worked, and we were on our way.

In the 1891 bylaws, the NYBG governing board is called, by a rather curious misnomer, the board of managers. This is true of a few other late nineteenth-century New York organizations, but we were bothered by this incorrect description of the people who are actually fiduciaries, so we changed the bylaws a few years ago and renamed them the more standard board of trustees. When I arrived at the Garden, James L. Ferguson was chair of the board. It was Jim who had hired me. Then the recently retired CEO of General Foods, his interest in the Garden was economic plants and the connections between agriculture and botany. But he and his wife were in the process of moving to Charleston, South Carolina, which was difficult for me because it was leaving me without a chair, and I was sorry, because I had known Jim for a few years and liked him.

But move the Fergusons did, and my move was to ask Thomas J. Hubbard, an active long-term trustee, if he would consider chairing the board. Tom was a person of high integrity and had the respect of many of the older trustees especially. He was the principal at a small private law firm, and he and his wife Anne were in a position, I suspected, to be very generous. Over lamb chops at one of his old-fashioned men's clubs in Manhattan, much to my surprise and relief,

Tom said he would be delighted to serve. He was very attached to the Garden, had been active for many years, and he thought it would be fitting. The nominating committee was easily persuaded. Tom's tenure lasted nine years, 1991–2000, a period of dramatic growth and change in the institution. He was tremendously supportive of me, and he and Anne became my best friends. Tom was interested in the institution as a whole entity—including financial management and rebuilding the physical plant, and Anne was intensely interested in field biology and the science program. They decided many years ago to begin to endow a series of distinguished scientific positions, and now, thirty years later, they have created fifteen endowed chairs in science, the herbarium, and the library. They never stopped caring about the institution. Tom has died, but Anne and I still talk on the phone regularly.

At the time Tom was made chair, the other most active, one might even say strongest, person on the board was Beth Straus. Beth had been a volunteer in the herbarium as early as the 1950s, and so she really knew the institution. She was also a passionate gardener. Although she understood the importance of the botanical research programs, her personal interest was on the horticulture side. But her true gift was in institution building, and because she really knew her way around New York, she was very skilled at recruiting new board members. Her greatest coup had been bringing Enid Annenberg Haupt onto the board in the mid-1970s. They were social friends, and Beth understood how much Mrs. Haupt could do for the organization. (And she helped me with Mrs. Haupt all through my early years.) Beth knew botanical gardens in America and Europe, and she knew the individuals who sustained those organizations. Her outreach was really something to admire. She introduced me to many distinguished British and Irish gardeners and landscape architects and to many New Yorkers who she knew could come to care about NYBG.

A valuable trustee Beth Straus had discovered was Ken Roman, at that time CEO of Ogilvy & Mather, the international advertising firm. A young, brilliant ad man who loved his Manhattan rooftop garden, Ken had talents and enthusiasm that Beth had seen could be very useful to NYBG. He was on the board when I arrived and had been on the search committee as well. He was a businessman and a creative one, and we thought alike. We immediately became friends, and he helped me codify my early thinking. Today, Ken is a Life Trustee, still active and constructive in Garden business. He has always believed in the high stature of NYBG and focused on the maintenance of that stature through audience development, marketing, and outreach. His personal interest has always been in our home gardening educational program, and he is one of the many trustees who have really good gardens of their own.

Many other trustees were tremendously helpful as we wrote strategic plans, devoted ourselves to institutional development, and conducted large-scale fundraising campaigns. Several were especially active. One of these was William C. Steere Jr., chair and CEO of Pfizer. His father, William C. Steere Sr., had been a specialist in the world of mosses, a respected teacher and mentor to a whole generation of American botanists, and president of the Garden from 1958 to 1973, so Bill Jr. had grown up in this milieu and has always been loyal to his family's legacy at NYBG. Through the 1990s, he was an extraordinary fundraising partner. He saw to it that Pfizer gave generously, and he asked many other corporate leaders for large-scale support, masterminding our annual corporate fundraising event for many years. He was also excellent at generating large-scale city and state funds for our capital needs. He and I had great fun working with the Giuliani and Pataki administrations. Governor Pataki, in particular, loved NYBG, as he had brought his children to the Garden when they were little, and he knew the organization well. We benefited greatly from the monies pledged by Pataki and Giuliani to Garden projects, and Bill was the connective tissue between NYBG and these elected officials. He and his family also helped us create and have now endowed our new state-of-the-art herbarium.

Also especially active among our trustees was Arthur Ross. Financier and philanthropist, dendrophile and passionate lover of parks and gardens, Arthur had been on the board for many years when I arrived. His principal interest at that time was the development through hybridization of a new variety of elm tree. He was deeply distressed about the loss through Dutch elm disease of the grand American elms that used to arch over American streets and shade our college campuses. He funded years of research which ended up in the patenting of the *Ulmus parvifolia* 'A. Ross Central Park' Chinese elm cultivar, and he promoted their use all over New York City.

Shelby White had joined the board just a few months before I arrived. She loves botanical gardens and had created the Country Garden in the Home Gardening Center in memory of her first husband, Rodney White. With her second husband, the investor and philanthropist Leon Levy, Shelby was responsible for many projects at NYBG, including the Leon Levy Visitor Center, the Native Plant Garden, and the Mertz Library's Shelby White and Leon Levy Reading Room. After her husband's death, she established the Leon Levy Foundation to continue their philanthropy, and has been a generous donor to many other botanical gardens and nature preserves in New York and abroad.

One of the most active socialites ever in New York City history (and that's saying something), Patricia de Bary was complicated and maddening and amusing. A trustee when I arrived, she was clearly in charge of our social events—at least she thought so. She was elegant, and she was smart but, I repeat, complicated. She and I tangled at first, but I came to respect and rely on Pat because she was really very energetic about raising money. She ran the Garden's annual gala—known in those days as the "Spring Party"—with an iron fist. She controlled the invitation list (it was stored in index files under her bed at home); she controlled the choice of chair and vice chair; she controlled the arrangement of the tables and seating according to her personal hierarchy of who mattered the most; she controlled the flowers and the food; she tried to control all of us. It was quite a dazzling feat, and somehow she made it work. To fund her parties, she craftily lined up two terrific supporters, Honoré Thorne Wamsler, who joined the board along the way, and Shelby White, both very wealthy women who were always happy to support Pat. Not only did she raise money, she also elevated our position among the New York social elite. For three decades NYBG has enjoyed a very elegant image with its social events, and it's only fair to say that the concept behind it was Pat de Bary's.

Marian Sulzberger Heiskell was the longest-serving member of the NYBG board. Mrs. Heiskell, as she was often called, had been elected in 1969, taking the place of her retiring mother, Iphigene Ochs Sulzberger, a grand New York City figure and environmentalist who had been on the board for some time herself. Marian served as an elective trustee and later as a Life Trustee until the time of her death in 2019 at age 100. (We like to brag that association with NYBG is often the key to a long life.) Marian was the child of a most distinguished New York family, the Sulzbergers of the *New York Times.* During the course of her life, she was the granddaughter, daughter, wife, sister, aunt, and great-aunt of six consecutive publishers of that venerable newspaper. After the death of her first husband, Orvil Dryfoos, Marian married Andrew Heiskell, then chair of Time, Inc. They were media figures, and it was a power marriage, of course, but delightfully, Marian was the most modest of women. She had real charm, and she loved her grandchildren, birds, the Audubon Society, the parks of New York, the greenmarkets of New York (the first of which she established at Union Square when she was chair of the New York City Council on the Environment), and the New York Botanical Garden. At the Garden, her principal interests were in the education of New York City schoolchildren—she wanted them all to be environmentalists and scientifically literate, as her mother had before her.

When I came to NYBG in 1989, Marian had been on the board for quite a while, and, as we were friends from our earlier days together at the New York Public Library, she announced to me that she wished me well at NYBG, but that she was going to leave the board. She felt she had lost interest and had not much more to contribute. She was probably the most notable civic figure associated with the institution, so I prevailed upon her, and she decided to stay. Often over the last thirty years she has referred to that and frequently told me she was happy she had. With her deep civic and philanthropic experience and her modesty, she was a woman of wisdom and solid judgment. Over the years, whenever I was discouraged or felt we were up against something too difficult to handle, I would talk to Marian. It was a great honor to have spent these decades with her—for me and for NYBG as well.

In the early and mid-nineties, the board also included Gerard Piel, publisher and editor of *Scientific American;* Edward P. Bass of Fort Worth, Texas, an avid environmentalist; Mrs. Thomas H. (Jane) Choate, a beautiful New York socialite who epitomized "Old New York" and had been for decades loyal to NYBG; Gilbert C. Maurer, COO and creative force at Hearst, a champion of the arts with a keen interest in gardens of the world; and Roscoe C. Brown, president of Bronx Community College, a high-profile Bronx civic leader, and one of the Tuskegee Airmen, the daring Black fighter pilots who were critical in the liberation of Berlin in 1945. The board numbered about fifty in those early years, and many were willing and able to be active.

This was a powerful and rather glittering cast of characters. In addition, the Garden was fortunate to have some notable friends. Martha Stewart was an early advocate, as was Karen Washington, the astounding community activist and leader of the urban farming movement, who headed NYBG's Bronx Green-Up program. The neurologist Oliver Sacks was also an enthusiastic booster long before he joined the board in 2003. During his many years on the faculty of Alfred Einstein Medical Center, he had brought patients to the Garden for his self-designed program of horticultural therapy. The film *Awakenings,* adapted from his book, had been shot at NYBG in the early eighties with Robin Williams playing Oliver. A dedicated member of the American Fern Society, Oliver recounted our fern research in Mexico in his *Oaxaca Journal.* He served on the board until his death in 2015.

Staff, administration, trustees, and friends—a stunning, dedicated, creative team.

THREE

Diving In and Learning Science

Despite my years at the American Museum of Natural History, there was much for me to learn about research science at NYBG. Curiosity about plants, I discovered, has been an aspect of human inquiry for at least two thousand years—we know of botanical gardens and exploration in antiquity. Many early botanical gardens were established to research *materia medica,* the therapeutic properties of plants. By the later nineteenth century, plant studies had diversified, and pharmacology was only one of many areas of interest. Our mission at NYBG, as codified in the 1880s, has been to discover, study, and deepen our understanding of the entire plant kingdom. Beginning with collecting expeditions and creation of the herbarium, NYBG has developed global research programs to investigate everything from plant systematics—taxonomy and evolution—to GIS modeling of species to economic botany focusing on useful plants.

All of these initiatives coalesce around our primary goal: saving the world's plants. Human life depends on plant diversity—not only do plants provide all of our food, fuels, fibers, and 40 percent of our medicines, they also clothe the earth and provide the oxygen all animals require for survival. Today there are 400,000 known plant species on earth, and current estimates are that one-third are at risk of extinction. Climate change, population growth, overdevelopment, habitat destruction—addressing these crises is of utmost urgency. At NYBG, our research fuels conservation activities locally, regionally, and around the world. We provide scientific data that shapes public policy and decision making. Particularly in Latin America, we have helped expand public awareness about biological heritage and motivate political interest in critical efforts such as the prevention of deforestation. For example, we are part of a coalition working to save what is left of the formerly vast eastern coastal forests of Brazil. At the grassroots level, we have established programs to teach local people how to manage their forest resources in a sustainable manner. Our research also addresses other urgent questions, such as how to deal with invasive species in wild places as far-flung as the Amazon Basin and Southeast Asia, and how to protect and manage agricultural lands in our country.

Another area of research focuses on the ways plants can mitigate or adapt to climate change, learning which species are unlikely to survive as they move up mountainsides to find cooler environments and which can flourish in a wide variety of ecological niches. In one project, we discovered plants that have adapted to live in deserts where there is no measurable rainfall—this data shows the potential to transform desertified landscapes into viable places for growing biofuel crops. Other research into heritage species can be used to address food scarcity—for example, where did our edible gourds and squashes come from, and how might new varieties be bred to adapt to hot and dry climates?

Beginning with the work of our cofounder N. L. Britton in Cuba, the tropical plants of the New World—the Caribbean, South and Central America, and the southern United States—have been the primary, but not exclusive, focus of NYBG research. Significant work has been done in other areas as well, for example, a comprehensive documentation of plants of the northeastern United States, and research in Southeast Asia and the Pacific Islands. But today, NYBG conducts the leading program of research in tropical ecosystems of the Americas.

As the result of our work over so many generations in Latin America, we have developed deep and lasting partnerships with hundreds of scientists and dozens of major institutions in these countries. We have educated hundreds of Latin American students in our graduate programs in New York and with our university partners have granted doctorates to young people from Brazil, Ecuador, Colombia, Venezuela, and other countries. Our extensive collection of tropical plants brings a regular traffic of scientists from Latin America to the Bronx. On a trip to Latin America in the early nineties to learn about our work there, I was greeted with a ceremonial welcome by scientists and university presidents in Rio de Janeiro, Caracas, and Quito. The New York Botanical Garden is held in high esteem by these institutions and government agencies dealing with natural resources and education. Today, our connections with Brazil are especially strong—the noted botanist Rafaela Campostrini Forzza of the Rio de Janeiro Botanical Garden is a Distinguished Counsellor to the Board.

Britton's work in Cuba inaugurated a long tradition of research collaboration there. For years, we have been the only American botanical garden involved in a program of identifying the island's many rare and endangered plants. There are more endemic species in Cuba than any other Caribbean island, and although there are

Gregory Long and Meyra Fernandez, director of the Cuban Institute of Ecology and Systematics, 2012. Photo by Scott Newman.

many national parks and protected lands, the plants in those areas are not well known. After a long hiatus, Cuban authorities invited us back into the country during the Obama administration, and the invitation was sanctioned by the U.S. Department of the Treasury. Our travel to Cuba was again prohibited during the Trump administration, but we hope to be able to return soon. There is much to be done.

The beautiful idea that botanical research has such far-reaching and important uses in the real world has inspired me and motivated and energized our trustees and staff over the past thirty years. And a big part of what we do is make our research, which can seem arcane, accessible and meaningful to our lay audience, our visitors, and, of course, to our funders in both the private and public sectors. All of us in the leadership of American arts and cultural organizations must struggle with how we articulate the meaning and value of our work—it's our "product."

The NYBG herbarium and library are integral to the research process, which begins in the field. From my years at the New York Public Library, I knew of the Mertz Library's preeminence and came to NYBG with a fervent desire to support and develop it. And after riveting briefings by then-curator Patricia K. Holmgren, I fell in love with the herbarium, where millions of dried, pressed plant specimens are labeled with the collector's name and where and when they were gathered. Led by the Brittons, the founders understood that the great botanical garden they envisioned would require a rich and distinguished herbarium and library, and they began in the 1890s to assemble them.

Britton had assembled his own herbarium containing his and others' specimens, and he was an avid bibliophile. His herbarium and library, and many acquisitions made possible through the generosity of Columbia University, were the seeds of the magnificent collections we curate today. We also have many older specimens gathered long before NYBG was founded: twelve by Darwin, several by John James Audubon, and many other rarities that illuminate the history of science. The herbarium has been built not only by five generations of NYBG scientists but also through the adoption of "orphaned" herbaria from institutions that have decided to abandon them for one reason or another, through generous gifts of personal herbaria from scientists working in other institutions, and through exchanges with our sister institutions worldwide. In addition to pressed plant specimens, our documentary collections include bits of plant tissue embedded in silica gel to be sampled for DNA. Nowadays, data are maintained in traditional herbaria and in frozen gene banks as well.

Containing nearly eight million specimens, in scale and scope the NYBG herbarium is second only to the French National Herbarium, the most extensive collection of botanical specimens in the world. The research faculty at NYBG goes out into the world, into the wild, into places where the natural world is not well documented, to find new species and to reconstruct the evolutionary history of those species and of previously known species. What they find is deposited in the herbarium. In this process we learn what the wild relatives of our crop plants are, what plants will be more likely to survive climate change, what plants have medicinal value, what plants, especially if they have economic value, are closely related to other potentially useful species. This is the traditional study of plants as it has been conducted since antiquity, and herbaria have always provided the baseline data as to what plants lived where and when.

Over half of the collection is digitized, and one and a half million specimens are also imaged. Our C. V. Starr Virtual Herbarium makes all this material universally accessible. Although the Steere Herbarium is not a public facility, we mounted a special exhibition in the Ross Gallery, at the main entrance to the building, which interpreted herbarium history, purpose, and methodology, showcasing scores of interesting specimens. It was a great success—we found that people respond enthusiastically to the beauty of pressed and dried plants.

The Mertz Library is the world's premier collection of its type. Unlike other botanical garden libraries, it covers not only botanical science but also the history of botanical exploration, horticulture,

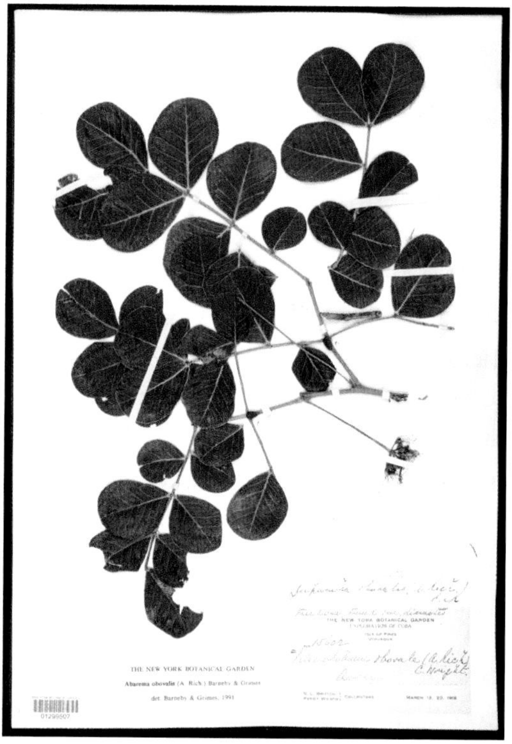

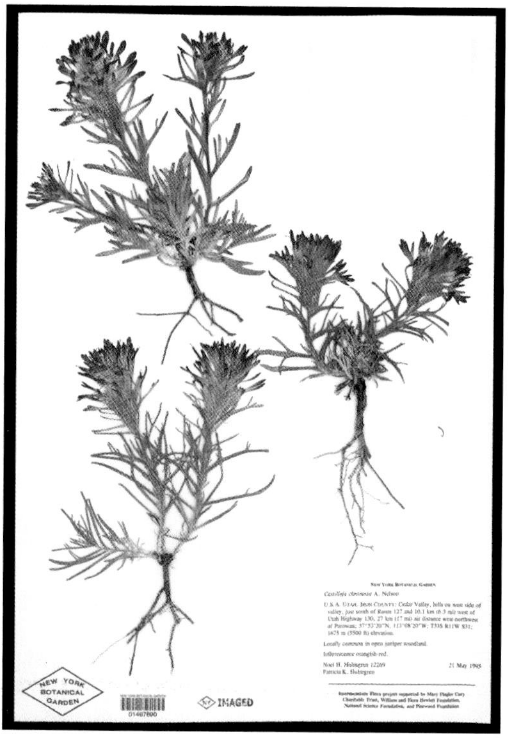

Two of the 7.9 million plant specimens in the Steere Herbarium: (*above*) *Abarema obovalis* collected by N. L. Britton in Cuba, 1916; (*left*) *Castilleja chromosa* 'A. Nelson' collected by Noel and Pat Holmgren in Utah, 1995.

landscape and garden design, and many other related subjects in the sciences and humanities. The library has been designated the New York State Library of Record for plant science, and because its establishment antedated the New York Public Library's, the NYPL has never felt compelled to collect comprehensively about plant science or horticulture. For 130 years, this task has been left to NYBG.

N. L. Britton was a visionary, and he was also highly entrepreneurial and a very good fundraiser. Two original board members, Andrew Carnegie, as a proponent of libraries as educational institutions, and J. P. Morgan, as a rare book collector, joined with him in a concerted effort to build a plant science library in New York that would rank favorably with the much older scientific institutions of Europe. In the years before the First World War, they established the Special Book Fund, with which Britton dispatched the librarian to Europe to buy rare and important books at auction and from dealers. Britton and his librarian knew what they were buying. They found fabulous treasures, and thousands of them at that, forming the core of the library's historical collections. In eighteenth- and nineteenth-century Europe, there had been a tremendous wave of publication of oversize, luxuriously illustrated books documenting science and garden design through those centuries. The Mertz Library holds a nearly comprehensive collection of them. It's every museum director's dream to be provided with plenty of money to acquire whatever will make the collections of their institution great, and Britton had the backing of two very rich and enlightened donors.

Over many generations, the library has also benefited from the generosity of private collectors. The early leaders of NYBG included Charles Finney Cox, a collector of first editions and autographed copies of Darwin's works, which are now among the treasures owned by the Mertz Library, and in our era major private collections were received from Elizabeth Kals Reilley (European garden history in rare editions) and David Andrews (a huge assemblage of rare materials documenting American botany and horticulture).

Often in institutions such as ours, research libraries are poor stepchildren, but the Mertz Library has benefited from substantial

Cedrus, from Christoph Jacob Trew, *Plantae Selectae* (1750–1773).

support of private donors such as Tom and Anne Hubbard as well as the Mellon Foundation and grants from the National Endowment for the Humanities and the Office of Education of the State of New York. About 50 percent of its annual operating costs are supported by income from restricted endowments, and long-term funding from the Mertz Charitable Trust is in place. The library has attracted this private and government support not only because it is one of the best science libraries in the country serving the academic community, but also because it is a public library—its research collections are available to people without academic credentials during many hours of the week. This is a long tradition at NYBG and one we are very proud of. About twenty-five years ago, when so many research materials in every discipline were beginning to be made available electronically, there was much talk about fears that libraries with real books and real archival materials would somehow become irrelevant—students and teachers would be doing all their research online. At NYBG we have never found this to be the case. Our beautiful books are used not just by scientists but by artists, graphic designers, students of the book arts, humanities scholars, publishers, and collectors. And many of our most current materials, available through electronic subscriptions, are accessible to our users who would find them too expensive to purchase.

Under the inspired leadership of Susan Fraser, now librarian emerita, the Mertz Library has conducted vast digitization programs and dramatically expanded and enhanced its primary function through an exhibition program and the creation, with Mellon funding, of a new division, the Humanities Institute, to engage fellows and visiting scholars in seminars, lecture series, and workshops and foster interdisciplinary research and public discussion. For the new exhibition program we built a small, beautiful gallery adjacent to the central rotunda on the sixth floor, as we had called for in our first strategic plan. Over the years we have held exhibitions of our illustrated books—on birds and plants, trees, the history of glasshouses—and medieval manuscripts and herbals. We also began to exhibit manuscripts and books from other collections (for shows

LuEsther T. Mertz Library exhibition gallery, 2002. Photo © Robert Benson Photography.

such as *Emily Dickinson's Garden: The Poetry of Flowers*) and works of art as well—Monet's garden paintings, the nature paintings of Frida Kahlo, Impressionist paintings of nineteenth-century American gardens, paintings from Georgia O' Keeffe's Hawai'ian period, paintings and textiles of the Brazilian landscape architect Roberto Burle Marx. Thirty-one of these popular exhibitions have brought hundreds of thousands of visitors into the library. Sometimes they would filter into the reading room and be invited into the rare book room, and consequently hundreds of thousands of people have come to understand what a treasured asset of NYBG and New York City the Mertz Library is.

In my education about science at NYBG, I had five mentors. Pat Holmgren, at that time acting vice president for botanical science, was also head of the herbarium. She and her husband Noel were also engaged in a large collaborative project to document the plants living in the geographic zone between the Rocky Mountains and the Sierra Nevada. (When the final volume of *Intermountain Flora* was published, the Holmgrens were awarded the 2012 Asa Gray Award for outstanding lifetime achievement by the American Society of Plant Taxonomists.)

Pat decided at the beginning of my tenure to take me in hand and teach me the basics of plant science at NYBG. This was a great event for me. She was a clear-minded teacher and loved her discipline. And she could tell what I was up to in terms of institutional change and approved of it. Pat and her colleagues in the herbarium maintained the highest curatorial standards, and she instilled in me a love and respect for the discipline—the meticulous work of the mounting room, where pressed and dried plants coming in from the field were attached to backings after they had been identified. She taught me to love the stories of early field trips and the historic specimens collected by Garden botanists in the late nineteenth century and also educated me about the significance of the research being done at NYBG, how essential it is to plant conservation and saving habitats. Pat championed NYBG's plant science, participating in international consortia and professional meetings and putting forward NYBG scientists for peer honors. She worried about the future of the institution, and she imparted that concern to me. I soon realized that she had made a determined effort to educate and inspire me. It was an invaluable service to the institution and to me.

In my early months on the job I also sought out four other mentors: Sir Ghillean Prance, former director of our Institute of Economic Botany and senior vice president for science, who had just left NYBG to become director of the Royal Botanic Gardens at Kew; the renowned Harvard biologist Edward O. Wilson; Thomas J. Lovejoy, the noted ecologist and environmentalist; and Peter Raven, the celebrated director of the Missouri Botanical Garden.

Iain Prance is a beloved figure in the history of the Garden. The energetic young Englishman had come to NYBG in 1963 as a research assistant, to conduct fieldwork in the Amazon. He was a romantic figure in the world of field biology and well known in international scientific circles. It was a severe blow to NYBG when he accepted the invitation of the Queen in 1989 to head Kew. He assumed that exalted position just a few months before I assumed the presidency of NYBG. I went to London to visit him at Kew, and learned a lot. Iain and his wife Anne had been at NYBG for decades, and they knew the score. They knew our science, but they also knew our financial problems, the personalities of the board members, the strengths and weaknesses of the scientific staff—everything. Over the years, Iain and I kept in touch. My experience in institutional leadership was in administration, financial management, fundraising, and strategic planning, and his areas of expertise were in scientific research. We could teach each other and always had a lot to discuss (although in retrospect, I'm not sure I helped him very much at Kew—he and Anne ran it with real verve). Handsome and dashing, Iain became a celebrity in Britain. But he was always generous to me in sharing his view of science-based conservation and his information and insights about NYBG, and in later years he enthusiastically supported our initiatives, both in science and institution-wide.

One day in 1990, I went to Cambridge to see E. O. Wilson, the unrivaled dean of American biological sciences. Wilson is an entomologist—what he really cares about is ants—but he knows all of biology, and he has a passionate interest in the well-being of plants. He's a fabled teacher at Harvard, and I knew he could teach me not only about biology but about all the interconnections of the natural world, the humanistic side of science, ecology, and conservation. I will never forget that day. Wilson was generous, and he clearly wanted to help me make a success of the new era at NYBG. His love of all of creation was infectious and highly moving.

I had known Tom Lovejoy somewhat over the years, and he was (and still is) a member of the board. He was trained as an ornithologist but had become a leading international conservation biologist—in 1980 he coined the term "biodiversity" and introduced the concept to the scientific community, predicting the dire results of climate change we are seeing today. He also has a gift for interpreting science for lay audiences. Tom and I became great friends, and he helped me articulate the value of research in solving or mitigating environmental problems. He guided me, over many decades, through many a sticky wicket, and he escorted me and some of our trustees several times to his famous LTER (Long Term Ecological Research) site in the central Amazon for serious trekking and a night's sleep outdoors in wonderful hammocks in the jungle.

Peter Raven, himself a highly respected botanist, had become by the late 1980s a messianic voice for the importance of species conservation. Larger than life, Raven was "preaching" all across America the importance of acting to save wild animals, plants, and places before they disappeared. I visited him several times at the Missouri Botanical Garden, which had been founded in 1859 and he had built into a botanical research and education center of international stature. He was intimidating—so insistent, so passionate, and somewhat chauvinistic about his fine institution—but, helpfully, he pushed my thinking in an interesting, conservation-focused direction.

Not only did these five generous people mentor me in learning about plant science, over the years they followed our work and helped me think through the growth and development of our research programs at NYBG. I also read as widely as time permitted about the history of plant research, especially in the New World. I believe I became an effective spokesperson to lay audiences on the importance of basic research in plant biodiversity, and in New York City philanthropic circles, in particular, our work has become better understood and our stature enhanced. That was part of my job, and I always enjoyed it.

FOUR

Rainforest Remedies, Complications, and Forging Ahead

When I had been at the Garden for a little more than a year, several of our senior scientists decided I should be taken to South America on an extensive field trip. The New York Botanical Garden had been working in Latin America since the 1890s and had developed expertise in plants of the New World tropics. After Britton's extensive fieldwork in the Caribbean early in our history, many NYBG scientists had come to be recognized world experts in the tropical plant families of Central and South America and the Caribbean. Our European sister institutions, although they had done some significant work in the Americas, had focused principally on Europe, Asia, and Africa. So, when we were established as an ambitious young institution here in New York, we gravitated to the Americas. By the early 1990s, NYBG had a distinguished roster of men and women exploring plants of the Caribbean and the tropical countries of South America—Brazil, Colombia, Ecuador, Venezuela, and French Guiana.

I had traveled as a tourist in Central America and the Caribbean, but I had never been to South America, and I had never been an ecotourist. My new colleagues thought it was time for me to travel with them to learn firsthand the excitement of plant exploration and to experience the hardships of traveling in remote tropical places—and the pleasure of it. My companions were Brian Boom, a specialist in the coffee family and ethnobotany, the study of how indigenous people use native plants; Michael Balick, head of our Institute of Economic Botany and one of the world's most respected scientists working in ethnobotany and ethnomedicine in the tropics; and Jim Luteyn, a specialist in the plants of the Paramo, the high, arid lands of the Andes. Our itinerary was to take us from Manaus in the heart of the Brazilian Amazon to the eastern cordillera of the Ecuadorian Andes and western Belize, a deep rainforest habitat where our scientists were documenting how the Maya had used plants for medicinal purposes.

We decided that December would be the easiest month to be away from our offices and other responsibilities for a couple of weeks—ignoring the fact that we would be traveling in the equatorial rainforest in the rainy season. I think that perhaps my three guides wanted to show me how tough the going had always been for them as they collected and studied plants in the wild. And they were successful. Except when we were in the high Andes, where it was dry and cool, I believe it rained every day. The temperatures were very hot, the humidity was of course very high, the mud in jungle paths and along trails was inescapable, and we sometimes slept in lodges with no glass or screens in the windows. It was pretty rough, but it was also pretty unforgettable.

Botanists in the field travel with permits to collect plants granted by the host country. And nowadays, we always travel with contractual documents stating that if any economic benefit is to accrue to NYBG, it will be shared fifty-fifty with the host country. Plant specimens, especially woody plants (trees and shrubs), are collected in roughly ten- to seventeen-inch pieces, preferably when the plant is in flower, in fruit, or in seed, as more can be learned from them at those stages. A scientist will usually clip about six to eight pieces of a plant and press them immediately in newspaper using a plant press carried on their back. The specimens are usually dried over a stove back in camp that night. When large bundles of these specimens have been gathered, perhaps after a few days, they are packed in cardboard or wrapped in paper and mailed from the nearest town home to the Garden. There, in the herbarium, permanent plant specimens are made, first by freezing them (to remove any insect infestations), then identifying them with their botanical name, and finally mounting them.

But back to the field. The scientists I was traveling with were collecting for our herbarium, and of course they were mainly interested in plants belonging to the families they are researching. But on this particular mission, we were doing something else very special. With the permission of our host countries, we were collecting bulk samples of plants for the National Cancer Institute, which was engaged in research to find potential remedies for AIDS and various forms of cancer. We had been contracted to collect thousands of plants every year for several years, for testing at the NCI laboratories.

Bulk collecting is different from collecting specimens. When Michael Balick and our team spotted a plant he believed indigenous people may have used in their pharmacopoeia, we would dig up a large portion of the root and take large sections of the leaves and stems and put them into an NCI bag. The Institute needed large quantities of material for their testing process. I helped Michael and his assistants fill these bags for many days in Belize. It was work that

José Mário Frazão (*left*), Gregory Long (*center*), and Cláudio Pinheiro (*right*) plant collecting in Belize for the National Cancer Institute, 1991. Photo by Michael Balick.

took tremendous energy and strength—especially wielding an ax to chop the root of a plant out of the rainforest floor.

Also in the 1990s, NYBG and other botanical gardens had contracts with pharmaceutical companies to collect plants to be screened for their potential as medicines. Forty percent of modern medicines are derived from plants, and many botanists and other specialists believe that there is much more to be found, especially in the tropics, where biodiversity is so profound. For Pfizer, Merck, and Bristol-Myers Squibb, we collected only stems, leaf tissue, and flowers—bulk collecting was not necessary for the screens they were performing.

At night, back in camp, we would eat rudimentary food and have a few drinks, as you can imagine, partly I suspect because we needed the sugar. This was certainly an eye-opening experience for me, which was exactly what my colleagues had intended it to be, and I confess now after thirty years that, exciting as it all was, it was also a tough two weeks. To make it all worse, in those days you could not fly directly from a Brazilian city to an Ecuadorian city to a Venezuelan city and then on to Central America. Every time we changed locale, we had to fly back to Miami and then out again to our new destination. And it was beginning to be Christmas time in the Miami airport—throngs of South American and Caribbean families traveling to their home countries, loaded up with immense packages of Christmas presents—appliances in some cases. If the lines hadn't been so long and so slow, it would have been funny.

In Belize, in particular, we often traveled long distances from one forested area to another in a Unimog, a massive heavy-duty truck made by Mercedes-Benz. Michael Balick had rented, or maybe borrowed, this monster machine, and it was quite necessary. The roads were entirely undeveloped—logging roads, often nothing more than a jungle trail—and no ordinary vehicle could have handled it. We had our most exciting adventure in it one day when we were many miles from a bridge across a wild river, but we were at a dam with a wide, roadlike surface on the top, which Michael believed we could drive across. But it was the rainy season and in our days in Belize it had never stopped raining. The river was roiling, and it was full of fast-moving trees that had fallen in from eroded banks. Maybe if it had not been the rainy season, the top of the dam would have been above water level, but on this day at least two or three feet of water was rushing across it. We stopped at the end of a muddy track and thought about the challenge. The river, by the way, was about 150 feet wide. I was terrified, and the other two scientists were not very happy with the prospect. Michael Balick is a fearless man, however, and he convinced us it could be done because of the tremendous scale and heft of the Unimog. He asked me, because I was his boss and the CEO of the organization, if I would give permission. Perhaps foolishly, I agreed that we should try, although I was pretty certain we would probably be swept into the water and pushed downstream by a humongous log. So Michael drove on, ever so slowly, and the Unimog beat the flood—we arrived, wet and scared, but alive, at the far side. I expect we drove directly to our camp and didn't collect many more plants that day. We had drinks.

This introductory experience was more than enlightening. It gave me enormous respect, of course, for the work of our field biologists. And it enabled me to organize many field trips over the next three decades for NYBG trustees and donors to see our scientists at work. These were a different sort of trip, though. My groups traveled when the weather would be fine; we found comfortable ecotourist accommodations; we cruised on the Amazon in a fairly luxurious boat; and we often began in Rio, where we visited the botanical garden, which is one of our principal sister institutions, and stayed in a high-end hotel on Copacabana Beach.

Among the many problems we encountered as we moved into the 1990s was one of great institutional difficulty. It was settled after intensive negotiations, and since then there has been peace and mutual respect between the protagonists. But despite that, I regret to say that in some minds there persists the idea that there is lingering distrust between NYBG and the Cary Institute of Ecosystem Studies. This is not at all true. Let me tell you the story.

It begins in the late nineteenth century with the tycoon Henry

Morrison Flagler. A founder of Standard Oil and railroad and real estate developer in south Florida, Flagler was the developer of Palm Beach and a principal figure in the Florida Keys. Two of his granddaughters were women whose names are important in New York social and philanthropic history. One lived in New York City and Palm Beach and was famous for her horticultural passion and expertise. In my lifetime, she was known as Mrs. Flagler Matthews, and she had a major garden and greenhouses in Rye, New York (belonging now, curiously enough, to a former trustee of NYBG). Another granddaughter was Mary Flagler Cary. She and her husband lived in New York City and were significant collectors of rare musical scores and antique playing cards.

Mrs. Cary loved country life and began to buy up old farms in Dutchess County, about eighty miles up the Hudson River. Over the next thirty-five or forty years, by purchasing many contiguous farms, the Carys created an estate of about two thousand acres in Millbrook, just opposite the historic estate of the Thorne family, who also have a multigenerational connection to NYBG. Mrs. Cary, who could easily have afforded it, never built a grand mansion on the estate. Instead she built just a small Japanese-style teahouse on a high hill with a pretty view of the surrounding countryside.

Mrs. Cary died in 1967. She had established the Mary Flagler Cary Charitable Trust, which owned the large landholding in Dutchess County. It was Mrs. Cary's wish that an institution of public education in environmental science and conservation be established there, and the trustees solicited proposals from various organizations. The New York Botanical Garden won the competition, and in 1971 the land was given over to the Garden, with some assurances that the Cary Trust would provide ongoing operating support. Over the next few years, the Mary Flagler Cary Arboretum was created, and NYBG recruited a horticultural and arboricultural team that began to plant collections for public education and enjoyment. As the 1980s approached, a small staff of ecologists was established as well, and a promising program of ecological research was launched.

In the early 1980s the leadership of the Cary Trust and NYBG decided to recruit a new head of the research program. They hired a highly respected mid-career ecosystem ecologist, who began to build an ambitious program in ecosystem ecology. Through the 1980s, there was certainly mutual respect between the botanists in the NYBG core science program and the ecosystem ecologists based in Dutchess County but not much collaboration. Ecosystem ecology and plant systematics are friendly sister fields, but they address completely different questions in the natural sciences. One of the major research programs of the Cary Arboretum, jointly sponsored by the Cary Trust and Yale, was based at Hubbard Brook in New Hampshire, where acid rain was being studied, an outgrowth of the research being done there on the chemistry of streams and nutrient cycles. So, the interests of the two scientific faculties were quite different, and with the exception of one or two projects, there was not much synergy.

This was the situation when I came to work at NYBG. In 1990, when I asked all the vice presidents of the institution to join with me in creating a bottom-up strategic planning process, all the vice presidents of the Garden were receptive and eager to participate—except the vice president of ecosystem studies at the Cary Arboretum. He had developed an independent program there and didn't see himself participating as an equal partner in a planning process headquartered at the NYBG base in the Bronx. This presented a problem for me and the trustees at NYBG because we respected his program and wanted him to be part of the family. But it wasn't what he wanted, and he soon lined up the leadership of the Cary Trust to support his position. And not only were they not willing to work with us, they decided to secede from NYBG. The Cary Trust, with massive funds available, was a great partner for the Arboretum—they were already supporting much of its work and covering the lion's share of its expenses.

The trustees of NYBG certainly did not want this separation, and so with the help of Tom Lovejoy and Ed Wilson, my office and the botanical science faculty in the Bronx tried to make a strong case for intellectual synergy between ecosystem and botanical science. Lovejoy and Wilson organized a distinguished symposium at the Garden in 1991 as part of the effort. But, in the end, we failed, and the Cary Trust and the Arboretum went their own way. The Cary Arboretum became a separate 501(c)(3) corporation and a supporting organization to NYBG in 1992. As a part of the separation agreement, the Garden was required to relinquish about $15 million in endowments created over the years by the Cary Trust for the benefit of the Arboretum; but in exchange, the Cary Trust made a $15 million grant to NYBG for botanical science, which we used to endow a suite of curatorial positions in the science programs. The Andrew W. Mellon Foundation made another major grant, meant to encourage us in the process of separation.

The negotiations were fraught, as might be imagined, because no one in the Bronx wanted the separation, but the Cary Trust intended to withdraw its massive support of the Dutchess County program if we didn't accept their request to secede. It was a difficult moment for us, as the Cary program was distinguished, and we valued it greatly. It was, however, not our core science and although it was well respected, it was considerably smaller and, of course, much newer than our historic plant science program. In retrospect, I have always felt that the loss of the Cary Arboretum in a sense did have a silver lining—it created a new and sorely needed $15 million endowment for our core botanical research programs; it eliminated an intellectual conflict that had been wrenching between the two scientific endeavors; and it freed us to focus on our traditional programs. But the struggle to prevent the separation and then to negotiate the settlement had been a terrible and enervating distraction.

In the early 1990s, it was clear to all of us that new and improved facilities for the library and herbarium collections were imperative, and this capital project became the central scientific element in our first strategic plan. The Beaux-Arts building at the top of the Tulip Tree Allée, opened in 1901, housed both the library and the herbarium—all of the collections, all the offices, all the reading rooms. It was terribly outdated and overstuffed. The herbarium was at that point around six million specimens, and although they were beautifully curated and carefully housed in steel cabinets, there was hardly room for a single specimen more. A new building for the herbarium was an essential need.

We knew that when all the specimen cases were moved to a new herbarium facility, the library building could be returned to the public rooms and rare book rooms that had been pushed by the burgeoning herbarium into a rear annex—much too small and inappropriate. So we made a plan to construct a large purpose-built herbarium building and to restore the library building, undertaking a program of adaptive reuse to create new public spaces, a reference room, a generous reading room, ample stacks for the library's general collections, a luxurious rare book room, and a suite of sunny offices.

We engaged Jim Polshek and Susan Rodriguez of the architectural firm Polshek Partnership. Together with our staff they devised a plan for a new herbarium building adjoining the library building, which would house all the specimens and accommodate at least twenty-five years of growth. We studied other institutions' facilities and tried to incorporate all the best ideas in ours—sophisticated climate controls, conservation elements such as filtration for particulate removal in air handling, materials with no off-gassing, and compact shelving to save space.

William and Lynda Steere Herbarium, 2002.

The planning, fundraising, and construction took nearly eight years to accomplish. But on May 1, 2002, we celebrated the opening of our new state-of-the-art facilities with great fanfare. It was announced to the assembled crowd, trustees, members, and staff that they would henceforth be known as the LuEsther T. Mertz Library and the William and Lynda Steere Herbarium to honor their major benefactors. The Steere family and the Mertz trustees were feted that day, as Senator Daniel Patrick Moynihan and Elizabeth Moynihan, both great friends of NYBG, cut the ribbon.

Of course, all of this building and restoration and adaptive reuse was very expensive—$65 million when all was said and done, which we had worked for many years to raise. Some came from the federal government through the Department of Agriculture and the good offices of our senators Moynihan and Alfonse D'Amato, some through the National Science Foundation, and some through the capital budget of the City of New York because of the enlightened thinking of the New York City Council leadership. The balance was raised from the private sector. The new herbarium building was made possible and endowed by the tremendous philanthropic support of the Steere family, William C. Steere Jr. in particular.

The new library would never have happened without the trustees of LuEsther T. Mertz. Mrs. Mertz, who died in 1991, had made a great fortune in the publishing business. She was never a trustee of the Garden and had been only a modest donor before 1990. She was known to only a handful of trustees, and I had met her only a few times before she died. But through her charitable trust Mrs. Mertz has become one of NYBG's two largest donors ever, ranking only with Mrs. Haupt.

LuEsther Mertz was a unique person in New York City philanthropic circles. An intensely private woman by nature, she did not fraternize with elite New Yorkers—in business, in society, or in philanthropy. She was the daughter of an educated but modest Pennsylvania family. She attended Syracuse University and studied library science. She founded the highly successful Publishers Clearing House with her husband in 1953. When her children predeceased her and her marriage had ended, she created a new family of

LuEsther T. Mertz with Mayor Edward I. Koch, c. 1985. Photo courtesy Larry Condon.

close friends and associates and named them trustees of her immense charitable trust, thereby arranging that her entire wealth would be given away.

In her wisdom, Mrs. Mertz had named wonderful people as her trustees, two of whom, Larry Condon and Bill O'Connor, joined the NYBG board and have led the Mertz philanthropy for thirty years. They were intensely loyal to her and continued her philanthropy in the style and spirit she had invented. They never used their privileged positions to make points, enhance their own status, or manipulate institutions. They were straightforward and honest and gave, always, with the utmost integrity. Just as she would have wanted them to do. The LuEsther T. Mertz Charitable Trust has now given away more than $300 million. At NYBG, the Mertz philanthropy currently amounts to nearly $100 million. In addition to NYBG, some of Mrs. Mertz's other interests were the Public Theater, where she was the first chair of the board back in the 1950s, the Natural Resources Defense Council, the ACLU, the Central Park Conservancy, and several environmental organizations, such as the Center for Conservation Biology at Stanford established by her friend Paul Ehrlich, who warned early on about population explosion and climate change. Mrs. Mertz was making climate change research grants as early as the 1970s. She was a visionary woman, a person who loved her garden at home but who also understood the impending world environmental crisis—fifty years ago.

What characterized Mrs. Mertz's philanthropy and continues to guide the thinking of her trustees and her closest partner, Larry Condon, is the notion that individuals and foundations with funds to give should listen to the people in the organization, find out what they most need, and provide generous support for it. In 1995, when we were laboriously raising the money we needed for the new library and herbarium, the trustees made a multimillion-dollar gift for the library side of the project. Mrs. Mertz had been educated as a librarian, and she was a reader, but the trustees did not make the gift because of her interests. They knew she would approve of it because she believed in the New York Botanical Garden, and this was our most pressing need at that time. Mrs. Mertz's giving was never about her—she would probably not even like the fact that we have put her name on the library. Her giving was about what her wealth could accomplish.

The graduate studies program of the New York Botanical Garden has a long and productive history. When last we counted, more than three hundred PhDs and master's degrees had been granted through the Garden's joint programs with several universities—Fordham, Cornell, the Yale School of the Environment, Columbia, New York University, and the Graduate Center of the City University of New York. This joint program between an active botanical garden and degree-granting universities is an unusual enterprise.

The NYBG graduate program is well known to faculty and students in the biological sciences at undergraduate institutions both in the United States and in Latin America, and they reach out to us. We help students apply to one of our partner universities, after we help them choose the most appropriate place for them—each university's plant science program has a different focus. Fordham is strong in plant ecology; Cornell specializes in molecular studies and traditional systematics; Yale, in molecular studies and forest ecology and public policy; Columbia, especially in the past thirty years, has focused on conservation biology; and NYU, on high-end plant genomics. CUNY's program encompasses all of these areas, and City University is more likely to admit our students coming from Latin America, for whom English is not their first language.

Over the past thirty years, we have had between twenty and forty students enrolled at any one time. A student's graduate committee is composed of our and their university's faculty. Students do most of their coursework at their university campus, although those courses are sometimes taught by NYBG faculty. Their research projects are conducted at the Garden in collaboration with our scientists. A typical example of how the program works is Chelsea Specht, a gifted young woman who had been an undergrad at the University of Delaware. When investigating graduate programs, she and her advisers contacted a member of the faculty of NYBG. She learned about our program, and we learned about her. She was intrigued by the combination of field biology and laboratory science being developed at that time by Dennis Stevenson, director of our laboratory programs, and his collaborators at NYU. We helped her apply to NYU, where she was admitted for her graduate work.

Chelsea was, like many of our students, interested in field biology because she had fallen in love with a plant family—in her case, the Costaceae, a pantropical family that includes wild ginger. We had faculty researchers working on similar families, and so, as soon as Chelsea had a summer off, she traveled to South America with NYBG staff as their graduate assistant (lackey). She was captivated with plant biology, with systematics or the study of the evolutionary history of plants, and with exploration. But at NYU she also was exposed to molecular studies, and she began to analyze the plants she was collecting in the field using molecular data. Chelsea finished her doctoral work with us, combining our field biology expertise and the newly available molecular labs at NYU. She graduated with honors and landed a teaching position at Berkeley. She is now the Barbara McClintock Professor of Plant Biology at Cornell. That's how it works.

Our graduate students like Chelsea have been a force in our

education as well. During the 1990s, they brought us into the world of molecular biology. Although we had traditionally been a field studies organization—our research grounded in collecting plants and studying their evolutionary history through analyzing herbarium specimens, we had also since our founding maintained an active lab program in plant anatomy, physiology, paleobotany, and phytochemistry. In the nineties, we learned that our students wanted to combine the new molecular techniques with our traditional research methods. Molecular biology was a powerful new approach to classifying plants and reconstructing their evolutionary histories, and it opened up a new universe of intellectual capacity. This awakening was happening at other botanical gardens too—Kew, for example—at the same moment.

Enter a visionary trustee of NYBG, Lewis B. Cullman, an investment banker and philanthropist who, with his wife, the television producer Dorothy Cullman, planned to give away their wealth during their lifetime. He had been trained as a meteorologist and had followed science as a layman, but what he really cared about was philanthropy—his own and that of other rich people. In his eighties and nineties, he waged a personal campaign to convince people to give away their wealth and advocated restructuring federal law to require private foundations to disperse more of their assets, a campaign that didn't make him very popular among many of our mutual friends. Cullman was interested in the science program at NYBG and also at the American Museum of Natural History. He was fascinated by molecular biology and to learn that at the genetic level plants and animals are not so very different—many of our genes are shared. This led him to fund a joint program in molecular studies at the two institutions. After a couple of years of struggling to create it, including a difficult time in recruiting a lead scientist, we were under way, and our graduate students were the first beneficiaries. The Lewis B. and Dorothy Cullman Program for Molecular Systematics has become central to the study of plants at NYBG (and animals at AMNH), and such programs now exist at other large botanical institutions worldwide. We have never abandoned field biology, but we have combined molecular studies with field studies in an innovative approach. Our graduate students led the way.

After the new molecular studies program was up and running, we knew that we would be interested in how plants function as well as how they are classified and how they have evolved. So in 2000 we established our plant genomics initiative and a year later formed the Plant Genomics Consortium with AMNH, NYU, and the Cold Spring Harbor Laboratory. The consortium has fostered robust and productive collaboration, with federal research funding over the years amounting to $25 million. Our molecular and genomics programs together are one of the most exciting aspects of twenty-first-century scientific research at NYBG.

FIVE

Serious Work on the Place: The Conservatory

Our great patron Enid Annenberg Haupt (1906–2005) cut a stylish figure in New York. She was the daughter of Moses Annenberg, the Philadelphia-based businessman and financier who owned the *Philadelphia Inquirer* and the *Daily Racing Form*. The family also owned *TV Guide* and *Seventeen* magazine—which she published. (She always said she wanted to teach young American girls how to be ladies.) Followed by her white Rolls Royce, she walked down Park Avenue every day from home to work. Mrs. Haupt lived in New York and Connecticut, had no surviving husband or children, and she adored orchids and her little dog, Orchid. Her orchid collection in her greenhouse in Greenwich was famous, as was her support of all the orchid growers' societies, the American Horticultural Society, the Philadelphia Flower Show, and NYBG.

In the mid-1970s, the massive Lord & Burnham greenhouse at NYBG was in terrible condition—it let in rain and the winter cold, which threatened the tropical plants; it was a disgrace. It was thought that it might soon be condemned. Then Mrs. Haupt attended an elegant fundraising dinner at the Garden, saw the condition of the building, and pledged a seven-figure gift toward its repair, in exchange for which the building would carry her name. The City of New York joined in to support the project as well, and the greenhouse was restored in the late seventies to everyone's delight. It became a well-known story—long before my arrival at NYBG. But by the late 1980s the building was already falling apart again, and it was my task to raise $25 million to restore it completely, from top to bottom. It was also up to me to secure major support from the city's capital budget and to ask Mrs. Haupt for another big gift—not even twenty years after she had funded it the first time.

There were three main players in the restoration this time: Mrs. Haupt; the famed Lord & Burnham Company, the builder of this magnificent greenhouse; and the City of New York and its capital budget. Lord & Burnham built most of the great American glass-houses of the era. The firm was based not far away from NYBG in Irvington, on the Hudson River. Although most of their grand Victorian greenhouses have disappeared or simply fallen down, a few fine examples still stand—the ones at Lyndhurst, Biltmore, the Phipps Conservatory in Pittsburgh, in Buffalo's South Park, and in Golden Gate Park in San Francisco. But at NYBG, we have always believed that our great conservatory, constructed in 1899–1901, is the queen of all Lord & Burnham greenhouses. Mrs. Haupt certainly thought so.

This crystal palace, 45,000 square feet (an acre) under glass, housing our important collections of tropical and desert plants, has been restored every twenty to thirty years throughout its history. In the 1930s, when it was only thirty or so years old, it was restored in a heavy-handed way—much of its classical detail was removed and replaced with a vaguely Art Deco look. The restoration in the 1970s was done by the respected New York architectural firm of Edward Larrabee Barnes. (At that time, Barnes worked with the noted American landscape architect Dan Kiley, who designed some of the surrounding gardens and flower beds, parts of which are extant today.) But, despite city and Haupt funding, the budget hadn't been sufficient, and the job was inadequate. By the late eighties the conservatory was in bad shape again.

When I began talks with the presidential search committee, there was much consternation on the board about how this new restoration could be funded. But even before I took office, two trustees and I went to work with the Garden's government relations department and succeeded in generating about $15 million from New York City toward the project. Now, the thing about greenhouses in northern climates is that in the winter it's hot and humid inside them and cold and dry outside, and as they expand and contract with temperatures that vary throughout the year, they start to fall apart. After all, they're only an eighth of an inch thick and mainly glass. When I explained, sheepishly, to Mrs. Haupt that the great conservatory bearing her name needed to be restored once again, she said to me in a very grand way, "Oh, don't worry, darling, I'm used to this. You have obviously never *had* greenhouses." Certainly not.

In spite of her cavalier remark, the ensuing negotiation was painful. Finally, after a long deliberation, Mrs. Haupt put in another $5 million and some new millions for an endowment, and we were set. When we had all the funding pledged, we applied to the city for a "pass-through" contract, which would permit our corporation to hire contractors and supervise the work—with some oversight by the city, to be sure—but without city agencies running the job. In the early 1990s, this was the state-of-the-art method of building and restoring capital facilities in cultural institutions (that are on city-owned land) with city funding.

This time the New York firm of Beyer Blinder Belle, who had a good track record in historic restoration, was chosen to design and help us oversee the project. The entire building needed to be taken apart and put back together again: 19,000 panes of glass, all curved according to dozens of arcs, had to be removed and replaced; tons of corroded cast iron had to be replaced with aluminum or another material; and huge supports made of tropical hardwoods, no longer sound (or politically correct), had to be replaced. After the design was completed, which took about a year and a half, and after contracts with the city were approved, also a lengthy process, step one was to close the building and remove all the plants. There were about four thousand valuable plants inside, both tropical and desert, some seventy feet tall, and they all had to be dug and stored somewhere for the interim. This process was masterminded by Francisca Coelho, a brilliant young horticulturist who would later become vice president for glasshouses and exhibitions. She and Garden hor-

Francisca Coelho, Merrin Vice President for Glasshouses, creating an orchid show.
Photo © Mark Pfeffer.

ticulturists dug plants with huge root balls out of the ground in the ninety-foot-high Palm House and moved them to off-site greenhouses. Some plants were too big to be moved and were housed inside wood and plastic constructions that stood like terraria inside the skeletonized building.

It was the most monumental task. As the construction work began, many parts of the building were stripped down right to the ground. There was one day in 1995 when Mrs. Haupt and I visited the construction site—when we had spent about $10 million on the project to date—and portions of the building weren't even there. We had battles with the city Landmarks Preservation Commission, because the conservatory is both a National Historic Landmark and a New York City Landmark. At one point, the commissioners insisted we replace dozens of hundred-year-old cypress window frames with reproductions created once again in cypress. It simply wasn't possible, and in the end they agreed to beautifully crafted aluminum. Architectural historians worked on the building, gifted horticultural designers helped us think about refashioning the exhibition designs when the building reopened, dozens of workers from many trades crawled across the frame of the building for many years, lead paint on the old metal members was ground off by workers wearing respirators and laboring over the hot summer in an air-conditioned tent erected over the building—it was all quite mind-blowing. This process was managed on the inside by a capital projects team we had specially recruited and overseen by a board committee of trustees who followed every detail with great concern and care.

Enid A. Haupt and Gregory Long visiting the conservatory in renovation, c. 1995.

The conservatory under a plastic tent during lead paint removal.

As if we didn't have enough to worry about, because this process was taking four years, we had to devise ways of letting the public know that NYBG was actually still open for business. In those years, when the landscape was still in such bad condition, most Garden visitors were coming to see the conservatory. It was tough for us to establish that the Garden could be open even while the conservatory was closed. To overcome this obstacle, we created an extensive program of public events—concerts, craft fairs, plant sales, happenings, family activities, and myriad other goings-on, and we marketed them vigorously to a broad audience to prove we were up and running and that their visit would be worth it. This program, running over four years, was budgeted at about $6 million, and we were thrilled that the LuEsther T. Mertz Charitable Trust, at that time a relatively new donor to NYBG, came forward to fund the whole thing—the events and the expensive marketing. As a result, we stayed in business, and during those years we even expanded our audience.

To prepare for reopening, Francisca and her horticulture team spent many months bringing the old plants back from their temporary homes in Westchester County and other far-flung places, acquiring new ones from tropical growers, and planting everything in a new design that everyone found very exciting. On opening day, May 1, 1997, the conservatory was thrillingly beautiful, with its new glass glistening on the curved surfaces, and all the plants inside happy to be home. Readers may not know that a glasshouse is too hot in the summer for tropical and desert plants—heat and sunlight are magnified by the clear glass—so every year it must be whitewashed in the spring to protect the plant collections from the burning summer sun. But on opening day, the seasonal whitewash had not yet been applied, and the building was a shining crystal palace.

We threw a grand party, inviting all of our patrons, high-level donors, elected officials, everyone who had worked on the project, and many of our Bronx neighbors—and 3,500 people accepted. On the most beautiful early May afternoon, before five o'clock when the party started, the streets surrounding the Garden were clogged with traffic. The parking lots were full, and guests had to park blocks away on the street—6,000 eager New Yorkers showed up! It seemed that everyone who had accepted our invitation had brought a friend. It was a real blowout. There was loud music, there were bagpipes, there were tents with food and drinks. The conservatory was resplendent.

Martha Stewart was one of the hosts, and she read a romantic poem to the throng under a big white tent. But around six-thirty, the brilliant sun was suddenly obscured by a heavy cloud bank. Inside the conservatory, we saw the darkness roll over, heard the thunder—and then the lightning began. Sharp bolts shot down, striking not very far from the building. It was terrifying and, of course, hugely exciting. I think all 6,000 of us were a little tipsy, and it was the most unearthly experience. After a few minutes the lightning subsided, the building had not been struck, and a torrential rain began. Thousands of people were jammed inside the building and the big tent. There was nothing to do but have another drink to wait it out. Of course, the white wine all disappeared, and I'm not sure anyone ever got their third glass.

At about seven-thirty, our trustees and many distinguished guests trekked through the pounding rain to the Garden Terrace Room in the new food-service building, adjacent to the conservatory, about 250 feet away, which was also making its debut that evening. Many people were soaked to the skin, but the dinner came off with panache. Believe me, though, when ten o'clock arrived, we were all happy to go home!

The reopening of the Enid A. Haupt Conservatory in 1997 inaugurated a new era of excellence and growth at NYBG. And it bears repeating that we could not have done it without the support of the City of New York and the people of the Bronx (whose tax dollars infuse the city's capital budget), the LuEsther T. Mertz Charitable Trust, and, of course, Enid A. Haupt.

In the opera houses of European and American cities in the 1980s and '90s, there was no greater star than Jessye Norman, and no more splendid voice. Raised in a church-going and gospel-singing family in Atlanta, Jessye Norman had achieved international stature and was revered for her dramatic soprano roles—Wagner, Strauss, Verdi, Puccini. She debuted at Metropolitan Opera in 1983 with a now-legendary performance of Berlioz's *Les Troyens* (when she performed both leading female roles), and in 1989 she was invited to celebrate the two-hundredth anniversary of the French Revolution, for which she sang *La Marseillaise* from a stage on the Place de la Concorde costumed in the French flag. Stardom was almost too small a concept for Jessye Norman.

Jessye, as we all called her at NYBG, was an active and passionate member of the board of trustees from 1995 until her death in 2019. We were deeply honored by her friendship. How did this happen? How did she come into our lives? It's a charming story.

In the early 1990s, while the conservatory was still under construction, we held an annual plant sale and auction in the spring to celebrate the new growing season and to raise funds. One year, attendance at the sale was low, as was the revenue, and the auction, which went on until too late in the evening, was more or less a flop. The plants weren't nice enough, and most of the generous donors who might have bought them anyway had gone home hours before. I was left with a few stalwart friends of the institution and quite a lot of staff among the small audience in the auction tent. The last big item to come on the block was an immense and very beautiful staghorn fern. Now, I love staghorn ferns, but at that time I was living in a small apartment in Greenwich Village and there was no place in my life for this magnificent plant. However, it was about to be sold for some ridiculously low amount, perhaps $150, and I, usually the underbidder, ended up with it. So, knowing that I couldn't possibly take it home, I decided that I would simply give it to NYBG, as we had many staghorn ferns in our collections and this was a beauty. It was a Friday night, and we were all tired. I went home and gave it not another thought.

Saturday morning my partner Scott Newman and I drove up to Otsego County, New York, where we had an old Greek Revival farmhouse as our weekend retreat from city life. We were there by noon on Saturday, and soon after we arrived I received a telephone call from a volunteer working in the Garden's new plant shop. Slightly embarrassed to call me at home, she asked me if the very large staghorn fern in the plant shop belonged to me. I told her it did, and then she explained that a woman had asked if the fern was for sale and really wanted to buy it. I replied that I was delighted, of course, and said, "Why don't you try to get $75 for it and we'll give the money to the Garden." I asked her if she knew who the woman was. She said no but whispered that thought she was a very important person. "Oh," I said, sensing an opportunity to make a new friend for NYBG, "can you put her on the phone?" And I heard the most marvelous deep dramatic voice with the most remarkable accent—it was Jessye Norman! I told her that I would like to give her the staghorn fern. She refused to accept that offer, so I said we would sell it for $75, and she said, it's a deal. Thus began her short journey to board membership and our long friendship.

Jessye Norman loved plants and NYBG. She had a garden she doted on, and she would often came to NYBG to shop for more plants and to ride around in the golf cart. An extremely generous person, she called NYBG "the People's Garden." For many years she attended the annual plant sale, which was combined with our Antique Garden Furniture Fair, a spring fundraising event, where she was sometimes the guest of honor. There would be four or five hundred Garden supporters there, and of course in those days not many people of color (maybe two?). Jessye would swing around with me to say hello to everybody and to look at every antique garden bench and every plant and have the most wonderful time. In addition to helping us out as our celebrity guest like this, she would occasionally accompany me to City Hall to speak to the mayor. Rudy Giuliani was an opera buff, and Jessye's presence at our fundraising meetings certainly guaranteed better results. Another time, in 1998, during a large-scale fundraising campaign, when she felt she couldn't give, she offered a Sunday night recital at Avery Fisher Hall, followed by a benefit dinner. She sang from her repertoire of Strauss lieder, and we raised hundreds of thousands of dollars. It was entirely her idea.

When she had been a trustee only a short time, Jessye gave

another great performance as host of the conservatory opening celebration. The event, full of speeches, including one by Mayor Giuliani, was scheduled for 11:00 a.m. Jessye came around nine o'clock for a walking tour through the conservatory with some of our horticulture staff. She relished learning about the plants, especially the tropical plants. She was keen to know about their medicinal uses, their traditional uses by indigenous peoples, their ecology, and their endangeredness in the wild. It was a wide-ranging and highly educational morning for her, and we all had fun teaching her and answering her perceptive questions.

At eleven o'clock, a crowd of four hundred was seated in the plaza in front of the Palm Court. Jessye wanted to open the program with singing "America the Beautiful" and asked everyone to rise. In unison, on the most beautiful May morning you could imagine, we all sang with Jessye Norman. I hope no one who was there that day has ever forgotten. Then she introduced the various speakers, who each spoke for a few minutes. At 11:45, the appointed hour for the arrival of the mayor, our principal guest, I was getting messages that he would be late, his entourage was miles away, somewhere in traffic. Since he was nowhere near, I joined Jessye at the lectern and explained that the mayor was delayed en route. So Jessye began to tell the audience about her morning in the conservatory.

Thomas J. Hubbard, Jessye Norman, and Gregory Long at the opening of the conservatory, May 1, 1998.

She remembered the names of the staff who took her through; she remembered many of the plants and their fascinating stories. When occasionally she couldn't remember a detail, she turned to me and I would answer. We continued this entertaining repartee as our staff kept slipping me notes that the mayor would be here in five, in ten, or maybe in twelve minutes. I began to think he wasn't coming at all, but finally he did appear, forty-five minutes late. He was thrilled to see Jessye Norman, one of his heroes, and made a very nice speech. But the mayor of New York was way outclassed by Jessye Norman.

A side note of interest to those readers who follow New York City politics: When the event was finished, formal pictures were taken. In the setup for the ceremonial group photograph, Jessye stood in the center with Mayor Giuliani on one side and Fernando Ferrer, the Bronx borough president, on the other. At that time the mayor was feuding with the borough president, who was a great friend of NYBG. One of the mayor's staff insisted out loud in front of everyone that I remove the borough president from the setup. Well, of course, I couldn't do that, and so I said to her, if you think that's necessary, you will have to do it. And she did. One picture was taken with the mayor and another with the borough president, and everyone went home pretty well satisfied.

Over the ensuing twenty years, Jessye's travel schedule and, later, her poor health kept her from being with us very often. But she had started a wonderful and important tradition—we would sing "America the Beautiful" in unison, sometimes with a borrowed opera singer, at the opening of every new garden and building. It's a tradition I hope will long survive at the New York Botanical Garden.

After we had begun to improve the Garden and upgrade the visitor experience, but during the time the conservatory was closed for renovation, we were all working mightily to market the Garden as a "perfect" place. We had an experienced horticulture staff, but in those days they weren't all very neat gardeners. As an avid tourist and lover of designed places, I had considerable experience in other significant gardens and landscapes, and our then head of horticulture Kim Tripp and

I were working with the staff to raise the bar. We never had enough money to hire all the gardeners we needed, but we were determined to overcome that problem. We were fussing with turf care, upgrading major gardens such as the Perennial and Rose Gardens, restoring the aging tree collections, and trying to beautify flower beds. With the help of some sophisticated gardeners on the board and garden designer Lynden Miller, who had been responsible for the Perennial Garden for a few years, we were beginning to see results.

Then one day in 1993, we looked up. In the air, rising above the conservatory, was a giant four-legged steel derrick. It had been erected in only a few days, and we'd been so preoccupied, we hadn't noticed until it was about two hundred feet tall. It was just inside the perimeter of the Fordham University campus, directly opposite the conservatory, which is on the edge of our property. So, these two huge steel-framed structures—one so graceful, the other so brutal, were facing each other, about three hundred feet apart. The thrilling westward view from the conservatory in the afternoon and early evening was gravely damaged by this hulking structure rising

Fordham University's radio broadcast tower under construction, c. 1998.

so nearby. And to make matters worse, it looked as if it were in the Garden—as if we had built it without thinking while we were on our hands and knees weeding flower beds.

I investigated. And I was told by officials at the university that the tower, which had not yet reached half of its anticipated 690-foot height, was intended to broadcast the FM signal of the university radio station. We were all angry and, urged on by several senior members of the board, I placed a call to Father Joseph O'Hare, SJ, president of the university. In retrospect, I believe that I didn't handle the conversation wisely. I was not diplomatic. Father O'Hare was an elegant and restrained man, and he was the president of one of the world's most respected Jesuit educational institutions. He wasn't especially nice to me, and he told me that several years earlier there had been a public notice in the *New York Times* that the tower would be going up—had we not noticed that? Of course we hadn't noticed a small public notice in the *New York Times* several years earlier. I asked him to take the tower down. He refused.

What to do? We certainly didn't want bad relations with our powerful neighbor, and we were well aware of the fact that Father O'Hare's political influence in New York City and New York State was considerable. He was chair of the mayor's influential Campaign Finance Board, and every time I went to Albany to talk to our legislators or agency heads about funding for NYBG, I would be surrounded by staff lawyers, most of whom were graduates of Fordham law school. But despite our concerns that Fordham would be too tough an adversary, we decided to fight. We wrote letters to the university requesting that the tower be removed. We hired lawyers—who explained to us that the university was building the tower "as of right." We immediately contacted the mayor's office, the Municipal Art Society of New York, the New York Landmarks Conservancy, the Preservation League of New York State, the borough president. Everyone was shocked; everyone was sympathetic; everyone loved the Garden and didn't want the tower built. Everyone wrote and agitated—Garden patrons, community organizers, schoolchildren, our Senator Moynihan and his influential wife Elizabeth, even Mayor Bloomberg. No one could do anything.

Early on in our campaign, we had succeeded in placing a front-page story in the *Times* with a ghastly picture of the tower looming over the conservatory. Partly as a result of the *Times* coverage, together with other lobbying and legal efforts (our legal bills of over $2 million were paid by the Mertz Trust), we were successful in getting a Stop Work Order issued, and construction was halted at about three hundred feet. There it stood until 2006, when finally Dr. Spencer Foreman, president of Montefiore Medical Center, located nearby, stepped in. The whole controversy had become an embarrassment for the Bronx. Foreman was a great institution-builder, having created at Montefiore one of the top hospitals in the Northeast, and he wanted the problem solved for all of us. He offered Fordham the top of a tall building on the hospital campus at the apex of one of the tallest hills in the Bronx. About this time Father O'Hare retired, and Joseph M. McShane, SJ, assumed Fordham's presidency. Father McShane and his trustees were eager to solve this long-running problem, and so they cooperated with Foreman. It was expensive to rent this space, and even more expensive to dismantle the tower and build a new one atop the high-rise building, but with generous grants (again from the Mertz Trust), the Garden covered all of those costs, and finally the tower was moved. It had been a long and costly thirteen years—during which we averted our gaze when we were in the vicinity of the tower—but thanks to Father McShane, Dr. Foreman, and the Mertz Trust, the situation was solved in the end, and the three institutions have been respecting neighbors ever since.

After the tower was down, Father McShane and I became good friends. He came occasionally to the Garden for lunch, and admired all we were doing to improve the institution. And I occasionally went over to Fordham to be taken into their notable buildings, including one from the 1830s, and to see their thrilling collection of old American elms. Although NYBG has a couple and there is a fine set in Central Park, I don't think any designed landscape has a better display of American elms than the one on the Fordham campus. They're revered by the leadership of the university and have been beautifully cared for over many generations. Cordial relations with

Fordham led to our developing various joint teaching and research programs in plant ecology and related subjects, and we now offer a joint PhD in plant science. We're proud to be partners in this academic enterprise with Fordham.

Postscript: During the years-long saga of the Fordham tower, Father O'Hare and I had an interesting encounter. Perhaps in year two or three, when he and I were barely speaking, our brilliant attorney Larry Lederman and I cooked up a scheme that we thought he might accept, which would result in the removal of the tower. To explain the idea to him, I invited Father O'Hare to lunch at the Knickerbocker Club, where I was one of the very youngest members. Rather to my surprise, he accepted. On a gloomy winter day, I went to the club, arriving there about fifteen minutes early. I passed through the central hall and greeted several good friends who seemed to be assembling before some kind of lunch party. Many of them knew, of course, about the Fordham tower issue, and they were highly surprised and titillated when Father O'Hare entered the building, passed through the hall, and mounted the elegant elliptical staircase to the small private dining room I had booked. I could practically hear the whispers and gossip running along the lower hall as Father O'Hare and I sat down for lunch.

Father O'Hare was at that time about seventy years old, and he was an extremely handsome man—a big globe of a head, beautifully groomed silver hair, shining black eyes, and a pink face above his white collar and his black suit. He was the perfect picture of a distinguished, elderly New York Jesuit. The room in which we were meeting was furnished with eighteenth-century English mahogany and wallpapered in a reproduction eighteenth-century French grisaille paper—a landscape entirely in gray and white and black. There was a fire on the grate, and the light flickered across the gray and white walls. Father O'Hare ordered a glass of red wine, which along with his face provided the only color in the room. I sat directly opposite him—behind and all around him were the gray and white scenes of the wallpaper, a backdrop to his black and white clothes and his pink face. It was a monochromatic portrait. I was deeply struck by the exquisite coincidence of the colors of this room and his colors, and although I was in the middle of making my offer about a potential solution to our problem, I couldn't help myself from blurting out, "Father O'Hare, I just have to say that you look perfectly beautiful in this room."

He didn't have dessert. He didn't accept my offer, and we never lunched again.

In 2018, I was granted an honorary doctorate by the university. Outside, under the elms, Father McShane presided. I actually felt rather sorry that Father O'Hare was not there.

John D. Rockefeller was a founding trustee of NYBG in 1891, and since that time the Garden has had a valuable relationship with this most philanthropic of New York families. During my tenure, the Rockefellers had been devoted to our Peggy Rockefeller Rose Garden, and in 2018, on my penultimate day as president and CEO, my assistant came into my office to tell me that David Rockefeller Jr. was on the telephone. Although I had known his mother and father over the years, I had never known David Jr. very well, and I was surprised to be hearing from him. He was calling with great news—as an executor of his father's estate, he wanted me to know that NYBG would be receiving a $5 million bequest from the Estate of David Rockefeller Sr. This was entirely unexpected! David Jr. couldn't have been friendlier, and he seemed to know that I would be leaving the Garden at the end of the next day. The bequest would be unrestricted, as most of the gifts we had received from his father had been, but it would come with the hope that it be used to benefit the magnificent Rose Garden. David Jr. told me that he and the other executors were proud to be making this gift to the Garden because of his father's long-term interest in the Rose Garden and because of his mother's interest in plants, the out of doors, and NYBG.

His mother, Peggy McGrath Rockefeller, had been an engaged board member in the 1970s. She was keenly interested in the preservation of American wildflowers—perhaps inspired by Lady Bird Johnson, whose Texas wildflower project she greatly respected—and

Peggy Rockefeller Rose Garden. Photo © and courtesy Larry Lederman.

had funded the NYBG publication of the fourteen-volume *Wildflowers of the United States* by Harold William Rickett. It was in honor of her birthday in the late eighties that her husband had funded the creation of the Peggy Rockefeller Rose Garden, a realization, finally, of the garden designed in 1916 by the noted American landscape architect Beatrix Farrand, which NYBG had never had the funds to build in its entirety.

I was very moved, of course, by this tremendous gesture from David Rockefeller Jr., especially coming as it did on the next to the last day of my decades at NYBG. I will always remember my visits with David Sr. and his philanthropic adviser Marni Pillsbury, when I would go to 30 Rockefeller Center every three or four years to ask for a new tranche of endowment for the Rose Garden. He often told me he had never felt that NYBG was one of his principal institutional responsibilities (those had been MoMA, Rockefeller University, and Harvard, among others), but that he did want to help us build the Rose Garden endowment. After Peggy's death, and almost every year until he was one hundred years old, on his birthday in June, a perfect time of year for seeing our three thousand roses in bloom, David would come and tour the Rose Garden and have a birthday lunch in the old Stone Mill overlooking the Bronx River.

All museums have physical plants, and many are old and suffering from various levels of "deferred maintenance," which is just a nice way of saying your predecessors have left you holding the bag. But not too many museums have surrounding grounds that are also their responsibility. At NYBG there was plenty of deferred maintenance in our 250-acre landscape, nine old buildings, and twenty miles of roads, sidewalks, and pathways, and we needed to rethink the maintenance and operations of the whole site.

The horticulture maintenance question was easier for me to wrestle with than the infrastructure question. I had a good sense of how our historic landscape should look and what kind of planning and labor it would take to bring it up to world-class standards. But infrastructure? No background on my part. Electric service? HVAC? IT? Sanitary systems? Stormwater management? All new to me, and on top of it, I am not a very technical person. But I believe that finding the right person for the task is a central demand of management, and that luck is an important element in life. And as luck would have it, we found Mark Cupkovic to manage our institutional operations and the restoration of our infrastructure. He came to NYBG in 1993 and stayed twenty-six years. He transformed the place: he was a godsend.

Engineers and operations professionals who manage campuses and big buildings know all about infrastructure, but museum managers and trustees do not. We had to address all the fundamental aspects of campus management in the late twentieth century, including saving energy and creating sustainability. (You can imagine what our carbon footprint was with an acre of tropical and desert habitats in a glasshouse 120 years old.)

And so Mark, in his beautifully methodical way, set out. First, the electrical situation: the service to most of our old buildings dated to 1958. It all had to be upgraded and redistributed and then expanded to serve the fourteen new buildings we built from 1995 to 2018. In this, Consolidated Edison, New York's utility company, collaborated with us in creative and thoughtful ways. ConEd's CEO at the time was a trustee of the Garden, and of course that helped. We decided our process should be first to repair or replace the existing electrical infrastructure and then to add to the Garden-wide system incrementally as line items in new capital project budgets. This made generating the required funding much easier, and by the time we finished, we had upgraded electric across the entire 250 acres.

Next we focused on HVAC and fuel (historically, our predecessors had heated with dirty, expensive No. 4 fuel oil). With supply from a city street nearby, we installed new high-pressure natural gas service and our own pumping station for our compressed-natural-gas vehicles—a hugely expensive improvement provided at a modest cost to us by ConEd. In addition, as part of a New York City "dual fuel" (natural gas and heating oil) campaign designed to reduce costs and emissions citywide, we converted all the boilers to dual fuel so we could leave heating oil for the last resort, on the coldest days of winter. We installed double-hull tanks underground. And we designed the "One Trench" utilities infrastructure project. We had to move gas, electric, fiber optics, and, later, water for irrigation, in a huge loop around the entire 250 acres, and Mark and his engineers figured out how to bury all four utilities together in one trench. The systematic installation of natural gas heating in all of our facilities, new and old, including large tents erected seasonally for special events, followed. We were getting modern!

The water supply question was addressed with the help of the city when, a few years later, after three scorchingly hot and dry summers, we recognized that we couldn't maintain our marketing position as "the most beautiful place" without irrigation. It was a difficult process—twenty miles of tiny trenches, snaking through the Garden everywhere? But it felt worth the terrific expense when, year after year, September–December were green and verdant, and those months became a robust period for visitor attendance and revenue.

Sanitary systems were next. The city had never installed sewers in the zone between the river and the Bronx River Parkway—with our septic systems, we were on our own! In the older Garden buildings these were primitive septic tanks, which, of course, were failing. We had to replace them with technologically advanced, highly engineered new tanks in the buildings located on the hundred acres east of the Bronx River. Some of these were massive, such the one for the new Nolen Greenhouses, a 47,000-square-foot facility with huge water drainage needs. For the remainder of the campus, we linked up to the city's inefficient CSO—combined sanitary overflow—an antiquated system that takes both sanitary and storm runoff to treatment facilities. There was some fancy engineering required, with the installation of hydrodynamic separators to aerate and clean stormwater and detention basins to hold it to reduce flow into the Bronx River. The basins at NYBG now hold 300,000 gallons. In all of this it was imperative that we respect the beautiful river, which cascaded down the length of our site. The state attorney general was watching, too—and we got flying colors for the rainwater filtration on the steep riverbanks and for other initiatives, including permeable paving, bioretention, large-scale sand filtration, and wetland overflow fields designed for the hundred-year storm, which actually occurred on April 15, 2007.

IT has been complicated, of course, with needs as varied as computers for four hundred office workers, high-tech equipment necessary for the new molecular and genomics labs, and the installation of a new computer cluster to serve the big data needs of our scientific research program. We also installed large imaging labs for the digitization of millions of plant specimens in the herbarium and books and other research materials in the library. Plus we upgraded the herbarium to fiber optics courtesy of the Garden-wide trench.

One of our last projects was the creation of an industrial-scale composting initiative we called "green materials recycling," and we constructed a several-acre "green zone" to accommodate it. We had always recycled leaves and other organic materials, even whole trees, but we were doing it without a proper facility or the right equipment. But at that time the Bronx borough president was Ruben

Diaz Jr., who was a progressive thinker and loved the Garden. He made a grant with which we acquired a mammoth Diamond Z tub grinder, which could take whole trees and turn them into chips and mulch. And with 30,000 trees on campus, the damaged or dead ones would provide a lot of mulch—millions of cubic yards. So now we make all of our own mulch, topsoil, and potting soil by recycling old plants. We designed a handsome concrete-paved yard with drainage and a stormwater management system to house this operation. After the new products are created, they're moved to a holding precinct with giant windrows, sometimes 15 feet high. State-of-the-art composting.

We were also concerned with energy use and greenhouse gas emissions. In this, our natural gas program was the most effective, reducing our carbon footprint by 50 percent. Often in collaboration with the city's Department of Energy Management, we made modifications to older equipment; installed reverse membrane roofs; converted heating systems from steam to circulating hot water; added highly efficient nanofluids in air-conditioning compressors to improve heat exchange (at a cost of $40,000 per gallon—compliments of New York City). We also installed occupancy sensors (why heat or cool to 68 degrees if the building is empty?), put reflective film on west-facing windows, and accomplished total LED lighting conversion. In the new Edible Academy, we installed solar panels, geothermal heating and cooling, stormwater capture, and composting toilets. Mark Cupkovic and his two engineers ran all of these systems from a centralized management system customized to operate many different types of facilities.

In 1989 we had inherited nine buildings. We built fourteen new ones to end up with twenty-three. We grew from 400,000 to 800,000 interior square feet. And our energy use and carbon footprint per square foot of built space never increased—they declined. The Edible Academy was certified as LEED Gold and several of our other new buildings had LEED certifications as well. And for years our collaborator in all this, the New York City Department of Energy Management, used Mark Cupkovic as a spokesperson for how old facilities could upgrade and become more sustainable. Mark retired from NYBG in 2019 to everyone's chagrin, but what a legacy he has left. He hasn't really retired, though. He has always been a classical musician—and his instrument is the cello. You should hear his Bach Cello Suites.

By the late '90s, our first plan was becoming rather famous and was providing the core curriculum for the strategic thinking sessions being taught at the Museum Leadership Institute, an intensive educational program for international museum executives held every summer at the Getty Center. I had been on the advisory board of the Getty Leadership Institute for several years, and Jeanne Liedtka, of the UVA Darden School of Business, was a member of the institute faculty. She was teaching about strategy and strategic thinking using a fairly worn-out case about Eastern Airlines. She had no museum material. Over the years of my association with the Getty, I had been telling her and the leadership there about our strategic planning process at NYBG, and finally, in 1999, Jeanne wrote the first of her three case studies about our planning process and its success.

At this time, I was also doing a lot of informal teaching in the Getty's summer program—informal because the academic leadership of the institute wanted Jeanne, the business school professor, to teach the concept of strategy and not "how to do" strategic planning. As museum managers, however, the participants were all desperate to learn how to do it, and so I was teaching them after-hours in rump sessions in the dormitories—practically under cover. I never understood why the people running the Getty Institute didn't want museum managers to learn how to fundraise, build boards, or conduct strategic planning exercises. They wanted to teach theory, not practical lessons. I am proud to be a practical practitioner.

SIX

A New Lab and an Emphasis on the Visitor

When our first strategic plan (for 1993–1999) was successfully completed at a cost of more than $207 million, all of which we had raised, we stopped and thought about what to do next. That we were a changed institution was clear. We were exhausted from the tremendous effort of raising the funds and executing all the projects we had assigned ourselves. So we took something of a break. But not for long. By early 2000, we were back in planning mode.

It was now the new millennium, and this time we launched the process with an expanded planning group and a new set of board members following the work. The planning office and facilitation were overseen by the new man on the scene, J.V. Cossaboom, our brilliant new vice president for finance. Once again, we structured the plan for a seven-year period, 2001–2007, and published it in May 2000. After the planning work was completed and many projects, programs, and desires had landed on the cutting room floor as we set our top priorities, the plan was costed out at $300 million, and that became our fundraising campaign goal. The components were the Annual Fund at $108 million—operating money we knew we would have to raise over the seven-year period whether we did a major campaign or not; new operating expenses of $36 million for initiatives above and beyond our current operating base; capital projects, $81 million—to be funded, we hoped, by several government sources as well as private philanthropy; and new endowment, $75 million (we knew this endowment goal would be a heavy lift—raising endowment funds always is).

Our priorities were, first, the operating funds we would have to raise year by year to balance the base or current budget; second, funding for existing deficiencies and new initiatives we felt were important—for example, new scientific positions, expansion of the horticulture maintenance team, and new IT developments. The capital projects deemed essential for forward movement of the institution were our third priority, and, finally, our fourth priority, endowment. Like many organizations, we were determined to reduce our dependence on private contributions for annual support. We intended to do this over our seven-year period by growing government support; by pushing to develop new programs that would increase earned income to the level of 25–30 percent of our annual revenues; and by raising new endowment funds, thereby increasing endowment income from 11 percent to 18–20 percent of our annual revenues. We also recognized that total annual revenue, to fund our basic program and the growth we anticipated, would have to increase by a total of 59 percent in the seven-year period. I believe, when all was said and done, that we achieved most of these financial objectives, and I know that we closed every fiscal year with a balanced budget.

The $81 million in capital projects was scheduled to fund seven important improvements in our facilities and landscapes. Since in our first seven-year plan we had done so much to improve our scientific facilities, of these new projects only two were aimed at them: to restore another third of the library interior—creating new rooms for public exhibition spaces, an imaging facility for herbarium digitization, and a new mounting room for herbarium specimens—and to build a new research laboratory.

Once again we brought in Polshek Partnership, who had designed the new herbarium building, to design the library interiors and the new lab. The new research facility, funded by New York State, the federal government, and a $7 million grant from Pfizer (orchestrated by Bill Steere Jr.) would be a major step forward for NYBG science. Designed by Polshek's Susan Rodriguez, the cream-colored build-

Pfizer Laboratory, 2006.

ing is long and low-slung, sited near and at a right angle to the herbarium, which is also cream-colored but high and wide. Together these buildings form two sides of a new science campus adjacent to and harmonizing with the historic Beaux-Arts library—creating a suite of structures at the top of Tulip Tree Allée. The open-format labs are lit by long skylights and floor-to-ceiling window-walls overlooking a lake and a quiet wooded corner of NYBG. Terraces provide outdoor space for faculty/student lunches. Housing our new molecular and genomics programs, our lab for structural botany, a freezer room for DNA storage, spaces for growth chambers and scanning electron microscopy, conference rooms, and faculty and graduate student offices, the Pfizer Laboratory was sorely needed as we modernized our plant research program. The facility serves not only our faculty and graduate students but also interns and visiting scholars. It is the perfect building.

It was while we were building the Pfizer Lab that the tragedy of September 11, 2001, shocked all of us in New York City. At nine in the morning on that sunny, hot Tuesday, we were all working away at our offices in the Bronx when news alerts began coming through on our computers and phones. We turned on radios and whatever televisions were available. We didn't know what kind of an attack this was, and we feared for our lives and our buildings—especially it occurred to me that the ninety-foot dome of the conservatory, so prominent from the air, might be vulnerable to attack. We closed the conservatory immediately; staff and volunteers were all evacuated. And then we asked everyone present at the Garden at that hour, about 250 people (it was of course before the public arrived), to come immediately to the Ross Lecture Hall. We announced that everyone would have to leave the Garden immediately. We asked people who had cars on the premises to move to the right-hand side of the hall, and everyone without transportation to gather on the left. We asked friends and people who might be neighbors to organize themselves into carpools for the evacuation, as we didn't think anyone should try to travel home on public transportation. It took only a few minutes for all of these employees and volunteers to form their groups and leave the Garden. By ten o'clock (and we still didn't know what was happening in lower Manhattan), the Garden was empty except for a few top security personnel, John Rorer, and me. We spent the day there like guards at the gate. Listening to the radio, in shock I suppose, but safe in our verdant landscape. Two days later, we were notified by City Hall that there would be major cuts in our municipal appropriation. But at the same time, we were all back at work and focused again on the forward movement of our organization. Shaken, of course, but devoted to our task.

In addition to the Pfizer Plant Research Laboratory project, there were five other major initiatives slated for this seven-year period, and these were all directed toward horticulture, exhibition development, and new programs to expand our audience to include culture-goers and to make Garden visitors happy. One particularly notable undertaking was the temporary installation in 2002 of sculpture from the Museum of Modern Art garden in the courtyards of our conservatory. How did this amazing opportunity come about? In an amazing way. Almost simultaneously three important figures in the lives of both MoMA and NYBG had the same idea. The museum was on the verge of closing for its massive expansion and renovation and planning to put the sculpture collection into storage for the duration. Scott Newman, my life partner and a partner in the architectural firm Cooper Robertson, was creating the master plan for MoMA, and two shared MoMA/NYBG trustees, Beth Straus and Lewis Cullman, had the brainstorm in unison: instead of putting it in storage, lend the garden sculpture to NYBG. A beautiful installation was designed by the landscape architect Susan Cohen and MoMA curators, and it all came to the Bronx—the Maillol, the Giacometti, the Rodin, all of it! It was a brilliant coup for us, and Glenn Lowry, director of MoMA then and now, and his trustees couldn't have been more gracious about it.

In a series of projects across the landscape, our plan called for considerable spending to make the Garden more beautiful. First, the irrigation system that has kept the Garden green ever since. Next, a restoration of seventy acres of landscape where there were two historic conifer collections. One was in the southern half of the property where in the mid-twentieth century our predecessors had planted hundreds of "ornamental" conifers—weeping hemlocks, dwarf varieties, and elegant nurserymen's selections—in a design by the noted American landscape architect Marian Cruger Coffin. This conifer collection is embedded in fifteen rolling acres of mature oak trees native to the site. Damaged plants were nursed to health or eliminated. Everything was fed and pruned. The turf was restored and new specimens were planted. New pathways were designed and installed.

The second large-scale undertaking was the restoration of a fifty-acre pinetum (a taxonomic collection of *Pinus* species, many grown from seed collected from the wild by NYBG scientists), surrounding the conservatory and planted over the past four generations. The collection includes not only pines but also firs, spruces, junipers, cedars, and hemlocks. Established in our early history, the collection had become sad and broken down by our day. We rehabilitated both the trees and the landscape, planted new trees, and opened up long views downhill to the conservatory. Then we installed interpretation and maps. The effort was generously funded by our dendrophile trustee Arthur Ross and his wife Janet, for whom the Conifer Arboretum is named.

The expense of these projects was well worth it. Because of the beauty of these tree collections and the high level of maintenance

Weeping Himalayan white pine in the Ross Conifer Arboretum. Photo © Robert Benson Photography.

we had established, the London *Times* reviewed NYBG and pronounced it "probably the best tended garden in the world."

Our ambitions for bigger and better flower shows in the conservatory dictated that we have an appropriate facility for propagating plants for exhibition. So we undertook the project of dismantling an old dilapidated complex of propagation greenhouses and pithouses and building a new facility. This huge new building is the kind of propagation and collection care greenhouse that every great botanical garden needs—a splendid behind-the-scenes facility backing up the conservatory. We spent years researching the best model, landing on a Dutch-style structure with a sawtooth glass roof. Designed by Mitchell Giurgola Architects, the state-of-the-art facility has seven different growing zones and an exhibition space housed in eight large greenhouses assembled in a unified 47,000-square-foot complex. Temperature, moisture, air circulation—everything is controlled by computer individually in each zone. Roof vents swing open when it's hot outside and close when it rains. High-tech, motorized shades move across when it's too sunny in the fern house. An evaporative cooling system cools the air. Sensors detect low humidity and hydro fans mist the plants.

Three of the growing zones are on the cooler north side of the complex, and in these we propagate tens of thousands of herbaceous plants—foxgloves, delphiniums, artemisias, and other flowers—for use in summer exhibitions, such as *Monet's Garden.* Also on the north side are rooms for our alpine plant collection and subtropical plants that require cool temperatures. The hot, sun-filled houses on the south side of the complex are for our tropical plants, including our huge orchid collection. Here we also grow and house the research collections used in our molecular laboratories. Altogether, when the production facilities are in full swing (typically December–August), there are close to 200,000 plants being cared for in this magnificent glass complex. The Bourke-Sullivan Display House is located at the entrance. Open to the public a few months of the year,

Nolen Greenhouses. Photo © Mick Hales.

this is where in some years we hold our spectacular *kiku,* or Japanese chrysanthemum, exhibition. Named for Wilson Nolen, chair of the board, in acknowledgment of a major gift, the magnificent Nolen Greenhouses for Living Collections opened in May 2005.

Many of our new facilities were now in place, and at this juncture we decided to think deeply and creatively about the visitor experience. In all public attractions, this is a challenging task and a critically important one. We want more people to come, and generate more income, and of course we want them to have a good time. We first devised the basic visitor experience in the early nineties and then rethought and revised it many times over the decades. As it is in most museums, working out the visitor experience at NYBG was akin to working out a 3-D puzzle.

The first questions museum managers have to ask themselves are, Who is the visitor? What do they want? How can we give them a good time? How can we encourage them to come back again? And most important of all, What do we want them to learn? We begin by thinking about the audience we are serving. As is the case with many large urban museums, we have a variety of audiences: culture-goers, elderly people looking for an outing, families with young children, professionals wanting information and professional support, tourists, and just plain neighbors. And they all need and want something slightly different. In a botanical garden, some need to ride, some want to walk, some plan to stay an hour, some want to stay five hours, some want to read Latin names on plant labels, some want to eat and shop, and some just want to relax or have fun. A main consideration is, how do the facilities developed by our predecessors serve the needs of these various groups?

At NYBG in the early 1990s, there were very few public restrooms, the main attractions (such as the conservatory and the Rose Garden) were far apart, the roads and paths were difficult to walk because the terrain is very hilly, and until 1997 there was no satisfactory public food service. Although we had inherited a beautiful historic landscape with vast and distinguished plant collections, a first-growth forest, and a rushing river with a picture-perfect waterfall, the Garden was not really a very visitor-friendly place. I discovered in my first few weeks, walking the paths to learn the lay of the land, that visitors had a hard time with finding their way because the roads and paths wind in and out among the trees, and the hills and huge rock outcrops prevent you from seeing ahead—all stunning, but all nervous-making when you are trying to get to the Rose Garden a quarter of a mile away. I also learned that there were almost no benches to sit and rest on—something we had to fix right away.

Within a few years, we had solved several of these problems. There were trams, each seating sixty, circulating in a logical loop around the Garden, stopping at the major attractions. An attractive new facility served lunch indoors or out. There were several new restrooms. There was more and easier parking than had ever been available. And many new benches. We also created a new gift shop—not as splendid as the one we later built at the new visitor center, but at least a gift shop. And there was the new Everett Children's Adventure Garden, which became a destination for more than 250,000 people every year, parents and children, teachers and school groups. But we were still reckoning with the difficulty of wayfinding in such a complex landscape and with how best to interpret our collections to educate visitors about the plants of the world. The considerations were complicated. For the most part, visitors are outside—there is the weather to contend with, the sun, heat, rain. Wayfinding and providing interpretation through digital experiences are much more challenging outdoors than indoors.

In addition, we knew that we absolutely had to have a facility for ticketing, orientation, and visitor amenities—we needed a "visitor center." A new entrance could be more expansive and welcoming than our historic Mosholu Gate at the Metro-North railroad station (which has always remained open, of course). One of our early master plans had identified a potential site on the western edge of the Garden, near the conservatory and the Children's Adventure Garden. This new site was centrally positioned on our north-south axis and directly opposite the attractive main entrance to our neighbor

Fordham University. And fortunately, there were no historic trees in the whole two acres—it was a tabula rasa.

When we considered the program for the visitor center, we realized that it had to accommodate many needs: not just ticketing and orientation, but automobile parking (outside the gate), group and membership sales, large restrooms for bus groups, locker rooms and offices for the visitor services staff, a retail shop featuring outdoor plant sales, and a café. Exclusive of parking, we would need to build about 35,000 square feet—it would be quite a large building. We began to talk to several architects, and our friend Hugh Hardy, at that point a principal in the innovative firm Hardy Holzman Pfeiffer Associates, came up with an ingenious solution. Instead of one large building, he proposed a cluster of five buildings surrounding an open courtyard in a loose asymmetrical design. The capital budget would be more expensive, but Hugh's idea, which we loved, was to greet the visitor with a long view into the Garden itself along a broad walkway down the center of the courtyard. The buildings, which Hugh called "soft modern," were all to be low, stone and glass with curving gull wing roofs, subsidiary to and reflecting the shapes of the surrounding hills and the dome of the conservatory only a few hundred feet away.

It was a brilliant solution. New York State bluestone was chosen for paving and some of the walls, inside and out. But the buildings are very open to the landscape, each with floor-to-ceiling windows framing views of the Garden. The courtyard, furnished with benches, trees, and containers of plants, is shady and breezy on a summer day and sun-drenched and warm, protected from the wind, on a winter day. It's entirely for pedestrians, too, as all cars have been left behind outside a high berm, which blocks the views and noises of the city beyond.

The Leon Levy Visitor Center opened in 2004, and it is a wonderful space just to be in out of doors, a piazza of sorts, enjoyed by visitors and staff alike. Hugh, who died in 2017, loved to visit NYBG, often with his children and grandchildren. He would sit on snowy days in the big café windows admiring the winter landscape. Many people think that among all of Hugh Hardy's distinguished

Leon Levy Visitor Center, 2004. Photo © Robert Benson Photography.

work designing and revitalizing theaters, museums, and other public spaces in New York City, this is perhaps his best. In a most self-deprecating way, he always said it was about the view and the Garden—he called it non-architecture.

So, what happens to the Garden visitor—let's say the "perfect" visitor—when she enters our premises via this beautiful gateway and set of amenities? First, she buys her ticket. Second, she buys a new annual membership and as a bonus is reimbursed the cost of the ticket she had bought just a few minutes before. This is because membership comes with a wonderful benefit, free admission to the Garden. Next, of course, since she's had a fairly long drive, she uses the restroom. Then she stops at the concierge booth to pick up a map, to learn about the new exhibition in the conservatory, and to find out which gardens are in flower. She learns that the conservatory is only a few hundred feet away by a direct walk that's easy to navigate, and she learns that the main stop of the tram that can take her around the Garden is only a few hundred feet in the other direction. Armed with all this information that will help her shape her day, she decides to stop at the café for a cappuccino. She sits outside in the sun with her coffee, admires the surrounding conifers, and realizes that she cannot see or hear the city outside. Without knowing it, she has left New York City, and she is about to have a day in the country.

She sets off. The show in the conservatory is about the gardens of the Alhambra. Filled with the fragrances of herbs, roses, and colorful citrus blossoms, the display enchants her. Leaving Spain, she wanders into the cactus-filled galleries of the desert plants of Africa and the southwestern United States—very different from the Alhambra garden—and she stops to think about the rich plant biodiversity of the world. Since it's a pretty day, she decides to walk to the Rose Garden. Heading down Azalea Way, she passes the calming pool of the Native Plant Garden, the entrance to the Thain Family Forest, and the entrance to the Maureen K. Chilton Azalea Garden (the azaleas seem not to be in flower, and she makes a mental note that June is not azalea season). After a ten-minute walk, following directional signage and her map, which is schematic and very useful, she ends up in the magnificent Peggy Rockefeller Rose Garden, thought by many to be the grandest and most beautiful in America. Since it's June, the roses are in full flower, and it is certainly the climax of her morning in the Garden.

After an hour in the Rose Garden, the tram stops just nearby and whizzes her in a big circle back to the visitor center. And since this woman is the perfect visitor, she decides to have lunch. She has soup and a sandwich outside on the terrace in the courtyard, reads some of the interpretative literature she picked up at the entrance to the Rose Garden, and calls her sister in Fort Worth to brag about New York City's cultural institutions. Then of course, she goes shopping. And being the perfect visitor, she buys two books in the NYBG Shop, which is the best gardening bookstore in New York, and she picks up a rosemary plant to keep on her kitchen windowsill. Exit, four hours later, the model visitor. So over the years circa 1991–2004, when we opened the visitor center, we worked out this experience for the adult culture-goer, we created the facilities she would require, and, with the right interpretation and guidance from garden staff, we helped her to have a successful visit.

But what of all the other different groups of people whose experiences we were concerned about? We guided families with young children to the conservatory but also to the Everett Children's Adventure Garden. The Adventure Garden is designed to teach, in an informal way, outdoors, about plant science and ecology. Visitors not wanting to walk all the way to the Rose Garden ride the tram and listen to the narration about Garden history and what's in flower. Our neighbors walk in, and because they are frequent visitors, they don't necessarily see the conservatory exhibition that day, and they usually don't shop. But, of course, they love the Azalea Garden when it's in flower, and the Rose Garden as well. These are a few examples of our thinking as we build and program to provide rich experiences for our public.

During the course of our second strategic plan, in the late 1990s, we were focused on marketing, and at one point we had an interesting conversation with the woman in charge of visitor experience at L.L.Bean's flagship store in Freeport, Maine. She pointed out that L.L.Bean and the New York Botanical Garden have something in common—we are both "destinations." Our visitors/shoppers must travel some considerable distance to get to our places—many people drive two hours to spend an afternoon shopping at L.L.Bean. Our visitors come from other boroughs, New Jersey, upstate New York, Connecticut, Long Island. The big lesson we learned from this interesting woman is that at L.L.Bean, they have structured a family experience designed to please several members of a family in many and various ways. The mothers who plan these trips are interested in shopping and in lunch. They are interested in keeping their children amused, and so there is a playground for the children. In the Christmas season there are horse-drawn sleighs—seasonal activities that are memorable and designed to give everyone a good time. She pointed out that a family planning an outing, such as a trip to

L.L.Bean or NYBG, has to devote the greater part of a day to it, and so it has to have aspects that please everyone.

We took this lesson to heart at NYBG and decided we could do even better than L.L.Bean at providing a variety of possibilities for our visitors: an outdoor environmental experience—hiking in the forest, for example; an educational or cultural experience—learning about Roberto Burle Marx's discovery of rainforest rarities; often, an art experience—Georgia O'Keeffe's paintings of the plants and landscapes of Hawai'i; many activities for the kids; good food; and shopping in our store or at our weekly farmers market. All this justifies leaving home at 11 a.m. and not returning until 5 p.m. And, of course, another benefit to us is the fact that this family will probably have two meals in the Garden, a coffee and pastry in the late morning and lunch a few hours later. Earned income is an important part of the revenue picture for us, and the family purchasing two meals is our favorite family!

One unique element in the visitor experience at NYBG is the great glass conservatory. Talking with people over the years, we have learned that everyone—men and women of all ages, little children—everyone is captivated by this spectacular building. Even children with little architectural experience or vocabulary are entranced by it. It sets the tone in the Garden, establishes a certain sense of old-fashioned elegance. So, as we designed new facilities—buildings, benches, interpretive signs—we kept in mind that the Garden has a signature elegant style that visitors respond to, and that style is an asset. We were thinking all the time about the quality of the experience for the people who are known nowadays in the museum business as our "guests."

Through our second very ambitious strategic plan, we were preparing ourselves to move forward in a robust way—restored landscapes, an inviting entrance, a superb plant propagation facility—all supported by vigorous outreach and audience development. When we had raised our goal of $300 million, we looked back at our accomplishments and were proud. We had envisioned where we wanted to be, and we were getting there.

SEVEN

The Board, the Money, and Michelle Obama

By the time we had finished our second huge fundraising effort and had established NYBG as a major player on the New York cultural scene, we had built on the cast of characters I inherited and created an even more powerful, collegial, dynamic, and younger board of trustees. I loved assembling this marvelous group. In most organizations, it is the CEO's responsibility to design a real team effort—starting with the board and the staff. The staff must be energetic and dedicated in all of their areas of expertise, and the governing body or board must include community stakeholders, philanthropists, people accomplished in the institution's programmatic areas, and government officials who care. And of course, it must be diverse. Blending together these groups of players, the CEO is usually (hopefully) the conductor of this beautiful orchestra, and it is often with the board that one begins.

Creating a great board is, on the one hand, a function of careful and judicious planning and research, and on the other, it is a kind of art. It is about making friends—the right friends, the friends you need to move the institution forward—and it is about institution building. Recruiting a wonderful board is a fascinating and rewarding kind of team building. Because you all have to enjoy being and working together, it's also like planning a successful dinner party. The mix, the balance, the tone all matter. You will want a rainbow coalition, an even gender balance, not too many big egos, not too many wallflowers, not too many businesspeople, not too many socialites. Lots of generous people, of course. No fools. Not even if they're rich. Some people who have achieved prominence in your field of interest. Some teachers are always good. Or maybe one teacher. An intellectual or two if possible—not more, because they will argue.

Not too many young people, not too many old people. I have always thought the perfect age bracket is about fifty to sixty-five. Younger people often have children at home, which makes their participation more difficult, and younger people are often not in a strong position to give money. And I think the American social system tends to create in middle-aged people an interest in community and giving back.

The NYBG board, with sixty-five members, is obviously large, but it functions beautifully. It needs to be big because in an institution with a multifaceted mission such as ours, there are many bases to be touched—we needed scientists, professional horticulturists, teachers, community representatives, social leaders, government officials, corporate leaders, and philanthropists. We have tried to keep the group balanced among all of these categories. Diversity has made our group much more interesting—diversity in race, gender, professional background, lifestyle. Something fascinating I have noticed over the years at NYBG and in the high-performance boards of other institutions I know well is that board members have often not known one another earlier in their lives—they have only met as they joined the group. I think this is a good thing, and it's very modern, actually.

In days of yore, boards of cultural institutions, especially, were seen as being like clubs—there was very little if any diversity, and many of the people knew one another socially or through their children's schools, for example, or their colleges or through family connections. They enjoyed being together because they were in the same social set. The old-school way of building boards was traditionally to tap a network of trustee friends. In today's world, I don't think this is the most successful strategy. The diversity I am describing results in the creation of a brand-new social set: what you all have in common is that you care about this institution and want to work as an ensemble to ensure its success and well-being. You share something much deeper than having gone to the same college. If board members are wisely chosen, and if a board is well managed by both the volunteer and the professional leadership of the organization, the bonding, the family feeling, will derive from the satisfaction of working together to realize commonly held goals. A board created and managed in this way becomes what I like to call an "assembled family."

At NYBG over the years, when a newly elected member was introduced to the board, generally very few of the board actually knew the person. The chair, various members of the nominating committee, and I as CEO had all met the prospect and discussed board membership with them, had vetted them and decided that they would add something valuable to the mix. Perhaps someone on the board had recommended the person and knew them well, but that was certainly not always the case.

We often identified new board members through their contacts with the development office or senior administrators. Generally, in searching for new trustees, we start with an analysis of people we already know and people who have demonstrated an interest in the programs of the institution. The process is initiated by staff, usually in the development or community relations offices because they know the most about our constituents, but a few key trustees do participate as well. Prospects may be generous donors, recognized scientists in our field, avid course-takers in continuing education, or civic leaders who have come forward to help us negotiate government funding. In some cases, a potential new trustee may have begun their association with the Garden as a Family Member at $100 a year or as an annual Patron at the $1,500 level.

Once we have identified likely candidates, development or government relations staff conduct research, and we begin a process that might be called "the cultivation of a friendship." With concentrated focus, this process will take a year or so, at least. It involves inviting the prospective trustee to various NYBG functions—grand beautiful galas and benefits, of course, but also lectures, tours, and community events, and at NYBG we offer lots of lunches and golf cart tours. I think over my twenty-nine years there were thousands of lunches. Trustees and senior staff are present at these events, and we all try to make it fun, to impress the guest with the excellence of our work, and to make friends. The depth of a candidate's interest is assessed. Sometimes a candidate is solicited for new or increased financial support along the way, and often this cultivation process will focus on the new friend's area of particular interest. Of course, it is important for the chair of the board, the chair of the nominating committee, and the president/CEO to be comfortable with the candidate.

It's interesting and important to consider what potential trustees might want for themselves from your organization—what's in it for them? If you can show the candidate that they'll find it, your chances of convincing them to join are greatly improved. At NYBG, some people wanted to learn to be better gardeners at home. Some wanted to help the people of the Bronx. Some wanted to feel instrumental in the global environmental movement—in our case, without being affiliated with an advocacy organization, without being on the front lines. Some wanted to travel with us to the Chelsea Flower Show or to Cuba on an ecotour. And some people, of course, want the social prestige of an old-line cultural organization. I always enjoy thinking this through and trying to give them what they want. In some cases, they learn about themselves—association with NYBG expands their personal horizons. If they gain, we gain.

Committee participation can be important to board development too. There really is no better way for prospective trustees to get to know the organization and for the organization to get to know them. There are many non-trustee committees working in conjunction with the development program at NYBG. They include the Patron's Committee, the currently active President's Fund Committee, committees organized from time to time to support comprehensive capital campaigns, a corporate committee for corporate fundraising events, committees of social leaders for every one of our social events, and a community relations committee. There are also programmatic committees not directly associated with the development program, for example, the Horticulture Committee (42 members), the Plant Research and Conservation Advisory Council (34 members), the Education Committee (28 members), and the Library Visiting Committee (31 members).

Over the years I have often been asked how I see the role of board members and how I define the relationship between staff and board. It is the responsibility of the CEO and their senior staff to include trustees in decision making and to make sure they are well informed at all times about programmatic activities, planning, and strategic direction. I believe that fundraising is a joint responsibility between staff and board. Fundamentally, the staff of an organization is responsible for management, and the board of trustees is responsible, in a fiduciary sense, for the financial well-being of the organization and for outreach. I think the best and most responsible trustees make an effort to be vigorous and knowledgeable ambassadors to the community. They should understand all of the programs and be able to talk about them compellingly. I do not believe that trustees should interfere in management matters. If a CEO is not trusted by the trustees, if the trustees do not believe that their CEO is capable of running the organization, they should find a new CEO. It is certainly the business of the board to find a CEO, but no board members should interfere with executive decision making.

Board committee structure, function, and management is a particular interest of mine. At NYBG there is a rather standard set of official, or standing, board committees—Executive, Finance, Audit, Nominating, and so forth. The Executive Committee is the most important. It is a high honor to sit on the Executive Committee, and at NYBG we have tried to keep it to about ten members. I believe that if the top committees are too large, they are hard to manage. Also, if a committee has a long list of members, the people not invited to join feel like second-class citizens. I have noticed occasionally at other institutions that more people are offended not to be included on a committee than are flattered to be asked to join—a real problem. Most board-level committees in major institutions such as NYBG do not need to be "working" committees. The

Investment Committee will need to be, often, but for most others, the members must follow the work of staff leadership, stay well informed, and participate in group decisions. They do not need to "work." In good governance, leadership from the CEO, the board chair, and one or two others (staff or board) is what matters, not groupthink.

The dramatic changes American cultural institutions have undergone in the course of my five-decade career have been driven in part by the need to fully engage the private sector, especially the philanthropic world, in new and compelling ways in order to generate more contributed income. At NYBG, and this is true for most American nonprofits, our sources of annual operating income are really only four: what we earn from business activities (tuition, selling tickets, membership sales, retail and food-service activities); what we receive from government sources; what we draw from our endowment—in our case, only 4.5–5 percent of the endowment's market value depending on the year; and what we generate in gifts and grants from generous individuals, corporations, and private foundations: fundraising.

To generalize about NYBG, our earned income and contributions are each about 25–35 percent of our revenue, and income from endowment and government grants is usually about 15 percent each. This is a pretty healthy spread, a "balanced portfolio," and although the percentages vary from institution to institution, and somewhat from year to year, many organizations would be happy if they were able to report this picture.

Over the past thirty years, our contributions needed for operating purposes grew from about $5 million per year in 1990 to $25 million in 2018. In addition, we ran four large-scale fundraising campaigns, based on the needs identified in our strategic plans, in which we asked donors for one-time gifts for various new initiatives, for capital projects, and to bolster our endowment strength—all to realize the goals of our strategic plans. As a result, the total we would need to raise in some years would increase to $50–60 million. During my tenure, we generated a total of approximately $1,200,000,000 in private gifts for all purposes and government grants for capital purposes. Of that amount, approximately $900 million was given by the private sector, an average over three decades of about $30 million per year. (But remember, we had started in 1990 needing only $5 million for the operating budget, and in those early years before our strategic plans were developed, we were not raising much capital or endowment money.)

How did we generate all this private support? Our campaigns were designed around our four strategic plans—they dictated our needs and shaped our strategies and solicitations. Because of our collegial strategic-planning process, everyone in the organization, from lower-level management up through the leadership of the board, understood our priorities and agreed with them. Of course, we had to maintain our large and committed board of trustees and a classically structured development office, and we all worked hard at it. Although only about 60 percent of the trustees were individuals with significant giving ability, they were wonderfully generous. They loved the Garden, and they believed passionately in the strategic goals we had laid out together. With the help of one or two professionals in the development office, I "took care" of the trustees, and together with the chair, asked them for their gifts. I also did most of the fundraising from major donors who weren't on the board. I am never afraid of asking high-net-worth individuals for major gifts when I believe in the cause and know that they do as well. Supporting me in major solicitations were the vice president for development (sometimes called "institutional advancement") and their staff.

I have always enjoyed fundraising—it's how you make things happen in the not-for-profit world. Asking people for money is not hard if you remember it's not for you—it's for the people in and served by the institution, and it's easy if you remember that it's part of a collaborative or community effort. You ask, they give. We all have our roles. Scientists save the plants of the world and, to pick another example, at the opera, musicians transport audiences. Fundraisers make sure there's money enough to pay scientists and musicians. Philanthropists do philanthropy. It's a group effort, and everyone plays their part.

In fundraising from individuals, it is always, of course, about relationships. I really like people, so creating and maintaining friendships was always fun for me, and I think that is true of all the best fundraisers. At NYBG we would try to maintain close relationships with perhaps 5–15 individuals capable of seven- or eight-figure gifts; 50–100 in the $50,000–999,000-plus category; 100–150 people capable of at least $25,000; and 300–400 individuals with capacity in the $2,500–25,000 bracket. You will notice that I write "capable"—these individuals were our prospects, from which our donors would be drawn by dint of their gifts. There is always a ladder of giving in a mature development operation—perhaps a donor begins as a $100 annual member, moving up to $1,000 in a year or two, then to $5,000, and then up to $15,000–25,000 after a few years, during which their interest in and affinity for the organization have grown. Donor relations, or "constituency development"—maintaining these relationships with several hundred promising donors and potential donors over many years—is a labor-intensive business.

Maintaining a corporate giving program is not entirely different—it is also about relationships. In the good old days, the relationship you hoped for would be with the CEO—perhaps they would be interested in joining the board? In the 1990s, NYBG was blessed, and we usually had eight to ten top New York CEOs on the board at any given time—Pfizer, Bristol-Myers Squibb, MetLife, J.P. Morgan, Chase Manhattan, ConEd, Mastercard, Texaco, Hearst. In the 1980s and until things changed in the early twenty-first century, I enjoyed my interaction and professional friendships with corporate leaders. We had a lot in common—especially a shared interest in working with teams. Profit versus nonprofit corporate life is not so different, except when it comes to motive. The funds a CEO could commit were not theirs; they were giving on behalf of their corporations to achieve a goal, often altruistic, of value to them. The funds I received were not mine—I had altruistic goals, too, of value to me. It was a nice way of doing business, benefiting both of us.

Nowadays, alas, things are greatly changed, and at least in New

York, it is almost impossible to engage so many top corporate figures. Sometimes their spouses are interested in being involved, but more often it is a professional in the company—the head of the corporate foundation or the vice president for public affairs, perhaps. It is still about relationship building in a carefully structured and managed program of engagement, affiliation, and stewardship, but in this case maintained by professionals in the development office. Often nowadays, companies like to make sponsorship gifts because they offer more marketing opportunities than outright unrestricted gifts—in this way they are seen as more valuable to the donor corporation. Think of J.P. Morgan's energetic sponsorship of TD Sports or MetLife's sponsorship of special exhibitions in art museums and botanical gardens.

Private foundations are cultivated for support as well, again through a process of making friends and building relationships, but it also involves the ability of the institution's professional staff to devise strategic linkages between the institution's needs and the foundation's interests. This is becoming more and more difficult in the current era as beleaguered private foundations hone their areas of interest down to very narrow programmatic guidelines, and program officers frequently direct their grants without much consultation with the supplicant—they are in control.

So, at NYBG over all those decades, we did it like that—as have many American institutions from the 1990s until now. One other major fundraising enterprise is "special events"—gala fundraising benefit parties. Our special events would sometimes raise $4–5 million per year, and they were killingly glamorous. It was part of our brand, and perhaps in the 1990s and for a while thereafter, we were one of the last vestiges of "high society." A loyal NYBG trustee might attend five of these events annually: the Rose Dinner in late September, when the Peggy Rockefeller Rose Garden was in full flower again after a sleepy summer; the Family Benefit in the conservatory in December with the *Holiday Train Show* as its centerpiece (bring your children); the Orchid Dinner, in Manhattan in February, if you were not in Palm Beach; the Founders Award corporate fundraiser in May, so often orchestrated for us by our friends at

Receiving line at the Conservatory Ball, June 2017: (*from left*) Maureen K. Chilton, Alexandra Lebenthal, Gillian Miniter, Catherine Waddell, Deborah Royce, and Gregory Long. Photo © Joe Schildhorn/BFA.

Pfizer; and finally, to close the New York social season, the Conservatory Ball, in early June. This fancy dress ball was the climax of our year: usually 500–600 guests for cocktails in the Perennial Garden, then dinner and dancing in the over-the-top decorated tent in the conservatory courtyard. Dinner was often provided by the fashionable New York caterer Glorious Food. Outdoors on a summer night, handsome men in black tie and gorgeous women in Dior, Oscar de la Renta, and Carolina Herrera ballgowns. The lucky ones in family jewels! Bill Cunningham, the beloved *New York Times* fashion photographer, would always document this party for his Sunday column, and everyone wanted to be in his pictures. The Conservatory Ball was high status in New York City, and it raised millions.

But like foundation fundraising, this method of generating funding is growing increasingly difficult. Even before the pandemic, in New York and many other cities, the field had become crowded, everyone deserving to be honored had already been honored (maybe three times!), and the social calendar had become very, very, very crowded. Parties do still work as fundraising events, however, and although we all wish the "no-party party"—when donors give but don't have to go—would work, it rarely has, at least not before the pandemic. But I guess despite our grousing about too many parties, we do still actually like them. Perhaps because so many email addresses are exchanged and so much real business is conducted during the cocktail hour. Or perhaps because so many people do still really like dressing up! As I write in 2021, I wonder what will happen post-pandemic.

When I entered the scene in 1989, the endowment fund of the New York Botanical Garden, after almost a hundred years in existence, had a market value of approximately $19 million (about $40 million today). This surprisingly small corpus was entirely inadequate given the scale of the Garden's operations. No one seemed to know exactly why the endowment was so paltry—large endowment gifts and bequests had been received from some members of the founding board and other generous philanthropists over the years, but because of either mismanagement or deficit financing, not much was left. It was dramatically clear to all of us in the leadership that generating new endowment would have to be a top priority.

This is the case in many not-for-profit educational and cultural institutions, of course, but raising endowment is always easier said than done. And raising the Garden's annual operating funds was already a daunting task during the years following 1989, when New York City drastically reduced the public monies allocated every year for the operations of its cultural institutions. We were having to fall back into ever more dependence on private sector fundraising. But somehow, over the past thirty years, we have succeeded in raising several hundred million dollars in new endowment.

How did we accomplish this? First, we simply asked donors to give "plain vanilla" endowment without restrictions, just because it's what we most needed for long-term security. But in slightly different efforts, we also created new endowed chairs supporting the salaries of distinguished staff members; solicited major benefactions that resulted in the naming of Garden facilities; and combined endowment gifts with funds contributed for capital projects.

When we began our program to build greater financial strength for NYBG, only two of the Garden's hundreds of positions had ever been endowed. Both were held by scientists in the plant research programs. So, as part of our comprehensive fundraising campaigns over the years we asked private philanthropists to create endowments permanently restricted to support various positions. For the first few years, almost all these were in the botanical science division because that area of the Garden's operation most closely resembled an institution of higher education and most private donors are familiar with endowed chairs in university departments—professors of English literature, deans of anthropology, and so forth. In addition, since many of the Garden's donors were knowledgeable about art museum fundraising, we were also able to remind them that curatorial positions are often funded by permanent, restricted endowments and named for donors. There was very little tradition of this at NYBG, however, and as far as I knew, none at all in other botanical institutions. But once again, we used the museum world as our model.

Whenever we created the outline of a new fundraising campaign, there would always be a section devoted to the generation of new endowed funds. As part of that effort, we would decide that perhaps ten or fifteen positions in plant science, horticulture, and education deserved to be endowed. They were positions that would always be necessary in the operation of the institution and that were currently occupied by well-known and well-respected professionals. A wish list might be Director of the Library, Curator of Economic Botany, Curator of the Rock Garden, Curator of the Tropical Plant Collections, Curator of Children's Education, and Vice President for Glasshouses and Exhibitions. Each of these was a central function, not optional or marginal.

We would then sort through our potential major donors to match their interests with the position. For example, a great friend of the institution was committed to our tropical and desert plant collections in the conservatory, and he knew and highly respected

Francisca Coelho, the vice president for glasshouses and exhibitions. We asked him for an endowment gift in the range of $1.5–2 million, promising that the senior curatorial position would be named permanently for him. Consequently, the position has been known ever since as the Vivian and Edward Merrin Vice President for Glasshouses and Exhibitions—and it will continue to bear that name no matter who is the future occupant.

We explain to donors that their endowments will be held inviolable on a permanent basis, that they will be managed by the Investment Committee along with all of NYBG's other restricted endowments, and that we will report to them annually on the activities of the person in the endowed position and the results of our investment strategy for the year. Like many institutions, we were usually striving to generate a 6–9 percent return on our endowments, but we were usually drawing for use in the operating budget no more than 5 percent of the market value of the corpus, thereby protecting our future buying power.

Most of our sophisticated and trusting donors understood this kind of arrangement because of their experience with colleges and universities or museums. There are now thirty-five permanently endowed positions at NYBG, created with gifts ranging from $1.5–5 million, depending on the importance of the position. And an interesting sidelight is that, over the years, the person holding the endowed position and the donor often became friends—they were partners, engaged in a common effort.

Another method for increasing institutional endowment is to ask donors of capital facilities to create permanent endowments to benefit the facility on an ongoing basis. For example, Mrs. Haupt adored the conservatory that had been named for her in recognition of her major capital gifts over the years, and in her spectacular bequest she left substantial funds to endow "her" conservatory. Unfortunately, not all donors have done this—it is often a great deal easier to persuade private philanthropists to build something than it is to convince them to endow it. Building is just more fun, perhaps more creative. And then, of course, some donors still don't believe in giving endowments to not-for-profit organizations. (I am afraid they think nonprofit boards and staff should suffer every year in assembling the cash we need.) So we have not always been successful, but we always have tried, and I believe that more than $100 million has been given to NYBG in this way over the past thirty years.

The Garden has also generated endowment gifts by asking major donors to "name" facilities or gardens and include a certain portion of endowment inside the amount the board determined would be necessary for naming. For example, a garden might have required for its permanent naming a $5 million gift, and the creation or restoration of the garden would have cost $2 million. The $3 million balance of the required contribution would then be used to create a permanent, restricted endowment to benefit that particular garden.

Regarding the concept of "permanently." It is true that the endowments are permanently restricted, as they are owned by the Garden. But, as the City of New York owns all the Garden's landscape and buildings, the city itself has ruled that facilities and landscapes can be "permanently" named for only fifty years from the date of the gift. As this is the city's practice, institutions such as NYBG now make this clear to donors wanting to name facilities in a long-term way. Facilities named before the rule went into effect in 2014 are grandfathered in and do not have such a limitation. In the past, the naming of buildings or landscapes in recognition of charitable gifts was generally considered to be "in perpetuity." In one well-known case, a major (high eight-figure, I believe) donor to a major art museum asked how long "perpetuity" was considered to be, and the director is said to have replied, "At our museum, fifty years." The gossip is that they settled on seventy-five. But this was years before New York City dictated fifty.

Endowment funds possessed by the private corporation of NYBG can of course be held in perpetuity, as the city has no standing with regard to the regulation of institutional funds. However, whenever possible, in contracts with donors we have always tried to insert the concept that if in the future the stated purpose of the gift no longer became necessary in the operation of the institution, the donor would permit the board to reallocate the endowment funds to a closely related activity. Most sophisticated donors have understood this caveat. With regard to endowments created for the funding of staff or curatorial positions, these are overseen by the corporation alone—the city has no role. And consequently, "permanent" in these cases means "permanent."

One especially enjoyable aspect of my role at NYBG was creating and leading luxury study tours for trustees and major donors. These were elaborate and highly sought-after perquisites for our best friends. Sometimes we took major-donor groups to the central Amazon, but our occasional trips to the Chelsea Flower Show in London, on which I would invite fifteen benefactors, were the top of the line.

In April of 1989, before I was to start full time at NYBG, I decided to make a quick trip to London to learn about our sister British institutions. I contacted several trustees of the Royal Horticultural Society and also my friend Iain Prance at Kew. I visited Iain and his wife Anne in their lovely Regency house just adjacent to the historic gate into the famous botanical garden. After a useful exchange of professional advice, Iain and his senior staff gave me extremely informative tours of the landscape, historic buildings, gardens, and scientific research facilities. It was an inspiring time for me—I felt a bit like our cofounders the Brittons when on their honeymoon they had said to themselves, "Why can't we have in New York a garden as splendid as this one?" And I determined to do what I could do at NYBG to build our institution so we would be perceived as a real peer of our venerable British sister—especially in botanical science.

I was also interested to learn about the work and programs of the Royal Horticultural Society, the premier educational organization in the world promoting horticulture through the operation of major public gardens, important publication programs, and a spectacular year-round series of London flower shows. Kew was known for its science, the RHS for its horticulture. A creature of the Victorian era, the RHS has been, for all of these years, largely responsible for the English love of gardening and the tremendous expertise of England's amateur gardeners. Although I had known about Kew and had visited it once or twice as a London tourist, I had no experience with the RHS. But I did some research, and I learned how far-reaching, powerful, and rich an organization the RHS is, and I wanted to learn everything I could from them that could help us at NYBG.

On a rather cold April day, I drove out to Wisley, the RHS garden in Surrey, to meet Robin Herbert, the president of the RHS, and Philip Browse, the director of Wisley. I always drive myself around the UK, usually without much difficulty with driving on the left-hand side, but I was a bit nervous that morning—I think I was insecure in my lack of knowledge about their organization and about really fine international horticulture. And of course I hadn't yet started as president of NYBG. When I arrived, early, at Wisley, which in those days had a rather messy parking field, an extremely modest visitor center, and rather substandard restrooms, I was feeling pretty unwell—even a bit sick. Maybe a little jetlagged and maybe just nervous about meeting the leadership of this distinguished organization. But by the time the appointed hour for our meeting arrived, I had pulled myself together, and my anxiety had turned to excitement.

The two men waiting for me in a lovely office in a large Tudor house were extremely gracious and welcoming and explained right away that they were very interested in the New York Botanical Garden and in what we might be able to do together as sister organizations. They said that in their many years of leadership at RHS, no one from NYBG had ever reached out to them, and although they seemed to know quite a lot about our institution (Robin had lived in New York at one point in his career), they said there had never been any interaction between their programs and ours. I was sorry, maybe even alarmed, certainly embarrassed, to hear that there had never been any outreach from NYBG, but I was delighted to learn that they were interested to become friends. Robin was chair of Marks & Spencer, a leading British department store, and Marks and Sparks, as it was then known, had just bought Brooks Brothers, so he was frequently commuting to New York, and of course I promised him tours of NYBG on his next visit.

They took me all around Wisley, showed me the inspiringly beautiful Rock Garden and exquisite Alpine Houses where their high mountain plants are displayed in gravel beds, flowering in April like jewels in Tiffany's windows. I saw the huge extent of the herbaceous borders, the walled gardens, the Rose Garden, the small gardens designed to appeal to home gardeners with modest properties. Then the huge beds, not yet planted for the season, where their gardeners conducted plant trials judged by committees of professional plantsmen—which will be the best new peony for home gardeners? which new delphinium will be the most beautiful blue? which asters will flower the longest in the fall? They were well respected for these trial beds, and are still today.

During our most cordial lunch, the subject of the Chelsea Flower Show came up. Taking place in London on the campus of Sir Christopher Wren's Royal Chelsea Hospital near the Thames, under a magnificent white canvas tent in late May, the Chelsea Flower Show was and is the RHS's marquee event, now famous to gardeners and garden lovers around the world. I had read about the Chelsea Flower Show but of course had never been, and I don't believe I knew anyone who had. They asked me if I would like to bring perhaps fifteen members of the board of NYBG to the show the following year. They said we should come on opening night, when in those days the Queen usually came to a reception to officially open the exhibition. The show would last from Monday to Saturday and would be attended by 150,000 eager gardeners, all of whom were members of the RHS. As they were making this dazzling offer, I thought to myself—"Well, *would* I like to bring fifteen members of the board to London next year to the Queen's opening of the Chelsea Flower Show? No question about that." I accepted this generous offer of friendship with alacrity. And as I left Wisley late that afternoon, I told Robin and Philip that I hoped there would be many opportunities for collaboration between our two organizations and that I would often bring American horticultural enthusiasts to Wisley in future years. Which for decades now I have done.

So, we went to Chelsea in 1990, and although we haven't been able to make it every year since—spring is a very busy season at

NYBG—we have attended a dozen times. On these trips, we have often hosted a festive dinner in London at the historic Brooks's Club in St. James's. We invite all of the Americans who have traveled with us, sometimes a few of their friends in London, and several dozen good friends from British horticulture—gardeners, journalists, authors, garden designers, and popular lecturers who have come to NYBG in previous seasons to present horticultural talks. We have always been very proud of our festive London dinner, the brainchild of Anne Hubbard, wife of my first chair, Thomas J. Hubbard.

Like several other New York cultural institutions, the New York Botanical Garden was founded on the concept of public-private partnership, and it still operates in that way. Created in the Gilded Age by the business and financial communities and by city and state elected officials, these institutions were understood to be cornerstones in the creation of a cultural life for the people of New York City. The first to be established was the American Museum of Natural History in 1869, soon followed by the Metropolitan Museum of Art in 1870, and then by NYBG and Carnegie Hall in 1891.

The concept was that the City of New York would own the land and buildings and the private, nonprofit, tax-exempt corporations with their individual boards of trustees would be charged with the management of the institutions and most of the fundraising necessary to support them. The collections and endowments of the institutions would belong to the private corporations. The original idea, although it was never a contractual arrangement, was that the City of New York would provide partial annual operating support sufficient to pay the salaries of maintenance and security personnel—in the case of the Garden, at the time I retired, that would have meant salary support for approximately 140 staff members, or about 20–25 percent of our overall operating budget. Over the years, however, the city's support has been eroded, administration after administration, to the point where now it covers only a fraction of these positions—when I last calculated it, the city's contribution to the operating budget of NYBG amounted to about 6 percent of what was required to run the organization. It has become a tremendous task for NYBG (and most of the other major institutions) to assemble over 90 percent of the essential operating funds from non-city sources.

There is one other component of city funding for institutional need, the capital budget, and this is a happier story. In many administrations over the past century, there has been some recognition that because the buildings of these organizations are owned by New York City, there should be some capital funds made available to the institutions to build (but not maintain) this cultural infrastructure. Over the past thirty years, in the best administrations, NYBG has received variously between $2 and $10 million per year for capital construction purposes.

But even the 6 percent or so and the capital grants are hard won. There is a group of now thirty-four institutions that operate in city-owned buildings, and we have been lobbying together for the past thirty years as the Cultural Institutions Group. For operating funds, we lobby as a unified force for increases in the city's allocation to the Department of Cultural Affairs, through which our funding flows. This is a very labor-intensive effort, which begins about six months before the city budget is passed. First, we lobby the New York City Council and then the administration or the mayor's office. On the one hand, it seems we have not been very successful, as the city's operating budget for cultural objectives has hardly increased in three decades. On the other, there are people who believe that we are fortunate to have maintained any city support at all. On the capital budget side, the method is different. We lobby individually, so you could say that we are actually in competition with our sister institutions for allocations from the city's capital monies.

Because of this history and all the government interaction required nowadays, NYBG, like all major New York City cultural institutions, has a dedicated government relations staff. They work year-round to keep our elected officials well informed about our work and impact. They also manage our relationships with state officials and agencies and, before the Obama administration, when federal earmarks were available, with members of Congress who could steer money in our direction. In the 1990s, NYBG had close relationships with our senators Moynihan and D'Amato, and with their help we were able to generate approximately $15 million in federal support, funneled quite naturally through the Department of Agriculture, for the construction of the new library and herbarium facilities. A few years later, we were also successful in generating substantial federal support for the plant research laboratories. Although the process was abused by many senators and congressmen (remember the "bridge to nowhere"?), for many educational institutions earmarks were a source of federal funds that enabled many worthwhile projects. It was certainly appropriate for the federal government to help us construct our plant research facilities, and the scale of this congressional funding was something we could never have achieved through applications to agencies.

Raising money, making friends with people who can help your organization, building a board, and positioning your organization so that influential people in government will support you—all of this requires the development of what I like to call "institutional marketing." It's related to marketing your product, selling tickets to the *Holiday Train Show*, for example, but it is a different effort. More strategic, not as strictly tactical. All through my career I have watched the cultural institutions of New York devise their marketing strategies, develop their positioning, create their "brands." It was my task at NYBG to identify the special character and attributes of the Garden, to package them, and to make sure that the staff and trustees and I communicated them in a unified message to our various audiences. I feel that the Garden's brand is its history and its accomplishments, and this brand exists at three levels—the international, the regional or municipal, and the local.

I believe that the high status and international recognition of the New York Botanical Garden date back to the establishment of the institution in the 1890s. The founders were scientists, financiers, and benefactors of international standing. Over the years since, the Garden's reputation for excellence, combined with high ambition, has enriched and underscored the international position of the organization. And, during our intensive period of growth and change, the efforts of the board and the staff to refresh, modernize, and improve the work of the institution added considerable luster to our position as a world leader among botanical gardens.

Our international status is mainly the result of the reach of our research in plant science, the scale and stature of the library and herbarium collections, the size and encyclopedic nature of the living collections, and the fine quality of our horticulture. The scope of our plant research and conservation programs has always been global, having focused for many generations on the discovery of plants across all of the New World. As the institution grew, botanical research grew as well, so that today the program is one of the largest operating in any freestanding botanical garden. The William and Lynda Steere Herbarium, at 7.9 million specimens, is now recognized as one of the two largest such collections in the world, and many professional observers recognize it as the best curated. The herbarium's facilities are also among the very best, and its digitization program is one of the top two or three globally. And its high level of activity (50,000 incoming specimens added and a similar number of outgoing specimens lent to other research institutions annually) makes it one of the world's most vibrant herbaria.

The LuEsther T. Mertz Library has now developed into the largest and finest library about plants in the world, and the Garden's publications program, founded in 1896, with its current issuance of several monographic works every year and three quarterly scholarly journals, has run uninterrupted for many generations and is depended on by academic scientists, land managers, environmentalists, and educators. In horticulture, international recognition has come as the result of the large scale of the landscape, the distinction of the topography, and the massive size of the living collections. As a physical and designed place, NYBG is recognized internationally as one of the greatest public botanical gardens. It has been designated a National Historic Landmark, and several of our buildings and landscapes are New York City Landmarks. Together, all of these attributes guarantee the position of NYBG on the world stage. The international significance of the New York Botanical Garden name is undisputed.

At the regional level, our brand is the result of our being one of six major New York City cultural institutions established in the late nineteenth century enjoying what has historically been a significant partnership with city government. The Garden's budget is the seventh largest of all the major cultural institutions in New York City, and the same goes for its staff, now numbering more than five hundred employees. Because of the illustrious names of the Garden's great nineteenth-century founders and because the board of trustees has always been composed of well known and distinguished civic and philanthropic figures, the institution enjoys a high social position in New York City. Our trustees, patrons, members, and even our 1.3 million–plus annual visitors are keenly aware of the Garden's role in the city, and they respect it.

Over the years, many members and visitors have told me that they enjoy the Garden's elegance and its history, and that being part of the Garden family is an honor and pleasure. Only a handful of other New York City institutions receive social or society media coverage as extensive as ours. There is considerable cachet accruing to individuals who are associated with NYBG. This sense of status is only one element, but it is an important element, in the Garden's fundraising and special events programs. When Bergdorf Goodman or Saks Fifth Avenue creates Christmas windows honoring major institutions in New York City, we want to be there, and we often are.

In our local community here in the Bronx, NYBG is respected and loved. Our local audience depends on the Garden's serenity and safety, knows that our high position in New York City enhances the Bronx, and understands that the international reputation of the institution is a great asset to the borough. Our extensive community service programs are treasured, and our local elected officials are proud of the institution. They see significant diversity in our audience and our workforce; they appreciate the business and economic well-being we bring to the borough; and they and their families love participating in the life of the Garden. The strength of our local reputation is evidenced by the financial support we receive from the city, particularly the funds directed here by local elected officials.

These three levels at which the Garden has developed fine

positioning are all interrelated, of course, and benefit one from the other. In addition, the Garden's great beauty, the escape from the city found here, the quietude, the plants and flowers and well-protected nature work together to create a sense of refuge. A kind of magical realm, the Garden has ineffable charm I like to say, and I think this charm underlies all of the brand development. We are about nature and art, about seriousness of purpose and education, and also about originality in programming and in the creation of the visitor experience. But on a practical note, achieving all this requires substantial funding for programming and for capital projects on an annual basis. In New York City, the excellence that underpins our brand at all levels is expensive.

All museums and other cultural institutions must concern themselves with their surrounding communities, and government officials and agencies at all levels are watching to see that they do it. The New York Botanical Garden is located in the north-central Bronx, and although the neighborhoods immediately around the Garden are mostly working and middle class, the fact is that 40 percent of families in the Bronx are living below the federal poverty level, and the Bronx is officially recognized as a "food desert," which means that fresh fruits and vegetables are hard to come by in local grocery stores. The Garden has long recognized that it has responsibilities to our neighbors and our community.

Many NYBG employees are residents of the Bronx, and some live so nearby that they can walk to work, which has been a great help to us, especially in the winter when there are miles of NYBG roads and pathways to be cleared of snow and the New York City streets are hardly passable. This is a long tradition at the Garden—there are people working here now whose families have worked for this institution for two or three generations. In addition to employing many community residents, NYBG operates educational programs in science and gardening for New York City schoolchildren. In many pre-pandemic years, ninety thousand children would come in their school groups for curriculum-related lessons in plant science, plant ecology, and organic vegetable gardening and five thousand city teachers would attend professional development courses at NYBG. These are two of the greatest contributions the Garden makes to its community. In addition, the Garden has free admission for New York City residents on Wednesdays. Our farmers market has been operating for many years on the free Wednesday. When I last saw the statistics, 40 percent of NYBG visitors were Bronx residents—the Garden is a valued community amenity.

Outside our gates, we operate an ambitious community gardening outreach program known as Bronx Green-Up. Initially established in 1988 to support volunteer efforts to clean up vacant lots and create vegetable and flower gardens, Bronx Green-Up runs workshops and training sessions throughout the borough, to help residents create community and school gardens and urban farms. Over the years, more than three hundred gardens and urban farms have been established through the program, which has become known worldwide for its impact on the lives of New Yorkers.

Yet another way the Garden relates closely to the community is in our fine relations with the elected officials who represent the Bronx at all levels of government—borough, city, state, and federal. We have always tried to maintain friendly and helpful relations with our elected officials, and for the most part, over my years as CEO, our local electives have loved the Garden and been grateful for all we do for their community.

In 2010 the Institute of Museum and Library Services awarded NYBG its highest honor, the National Medal for Museum and Library Service, in recognition of outstanding contributions to the community. At the award ceremony held at the White House, Michelle Obama read a stirring tribute to NYBG for our service to our community and, she even added, for our service to the international conservation movement. I was told to bring just one member of the board of the Garden with me, and so of course I asked Karen Washington, the active and influential leader of Bronx Green-Up, and a fine member of the board of trustees. We had a marvelous time wandering around the White House, being serenaded by the Marine Corps band, receiving our award from Mrs. Obama herself,

First Lady Michelle Obama with NYBG trustee Karen Washington and Gregory Long at the National Medal for Museum and Library Service award ceremony, 2010.

and having many photographs taken. In one group photo, Karen, who is all charm, asked Mrs. Obama if she could put her arm around her in the photograph. Mrs. Obama, who is also all charm, laughed and said that would be wonderful. After the awards had been handed out by Mrs. Obama, a festive and luxurious lunch followed in the State Dining Room, for us, other awardees, and museum friends from across the country. It was a great moment for the New York Botanical Garden.

Meanwhile, as NYBG was being honored in Washington, as trustees were giving and raising endowment, and as our elected officials were helping us at City Hall, the fine professional staff we had assembled and nurtured were daily operating myriad programs and activities. In other words, while all the planning, strategic development, capital construction, new programs, and innovation were undertaken, the strong core of the institution was also flourishing. Our landscapes were beautifully maintained by our horticulturists, arborists,

Thain Family Forest. Photo © Marlon Co.

growers, and curators of the living collections. The grounds were safe—no crimes against people in many consecutive years. The plant research faculty was highly successful in their discovery of new species, in their publications and teaching, and in the generation of substantial funding from the National Science Foundation. We were engaged in our ongoing educational programs for schoolchildren and families. We were graduating six to ten students every year from our School of Professional Horticulture. And our continuing education program was teaching plant-related studies to five thousand adults every year and attracting dozens of international lecturers to programs for as many as four hundred attendees. Somehow we kept all of this going while we expanded and moved into new areas. The people of the Garden, my fellow workers, deserve all the credit.

EIGHT

More Serious Work on the Place: 1.3 Million Visitors

Based on the objectives of our third seven-year strategic plan, we set out in 2009 to raise a fresh $480 million. This new effort was hard on the heels of the success of our second plan, which had exceeded its $300 million goal by $26 million. The board was energized. The staff was working together in a powerful, creative way. And our international identity as an important cultural and research institution was firmly established. We had tremendous momentum. Having accomplished so much capital construction in our second strategic plan, our focus in the third plan turned to landscape renovation and new programs linking the arts and humanities to gardening.

Over the years since its founding, the Garden has had several master plans for its large "cultural" landscape and the facilities needed to conduct its extensive programming. These plans have all built upon one another—enriching and elaborating on the core design precepts set out in the 1890s—and they have all respected the natural character and flow of the historic place. The first was created in the mid-1890s by the architect Calvert Vaux, codesigner with Frederick Law Olmsted of Central and Prospect Parks. At the time, almost four decades after Central Park was built, Vaux was working for the New York City Parks Department, and he was asked by the NYBG governing board to devise a schematic plan for visitor circulation and the location of buildings on the newly acquired 250-acre site in the Bronx. Soon after he was commissioned, however, Vaux became ill and died. Only one rudimentary drawing survives, which shows Vaux's early thinking. Interestingly, his plan reflects the old country roads and cow paths that had traversed the site during the nineteenth century, and some of which were still in place in the 1890s.

The scheme was taken up by John Brinley, the new institution's vice president for horticulture, and the well-known landscape architect and nurseryman Samuel Parsons Jr., who had been Vaux's protégé and succeeded him as head landscape architect of New York City. Brinley developed the circulation system and building locations that still exist today. It was during this era that both the Lord & Burnham conservatory and the Museum Building (now the library), designed by Robert W. Gibson, were constructed. Brinley also laid out the early collections, including conifers and herbaceous plantings. The fundamental design concept for the Garden was that the original hilly terrain with its sometimes gargantuan, whale-like

View of the Garden's rolling lawns and curving paths. Photo © Marlon Co.

rock outcrops of Fordham gneiss and its thousands of mature deciduous trees would be maintained and would be developed in an Olmstedian, picturesque style, with rolling lawns and curving pathways sweeping around natural features. The fifty-acre first-growth forest in the center of the site was not to be touched.

In contrast to the surrounding landscape, two monumental Beaux-Arts features were created—an immense rectangular platform raised a few feet above the surrounding lawns to provide a formal setting for the conservatory and a five-hundred-foot processional way of two parallel carriage drives, flanked by a double allée of American tulip trees, leading uphill to the Museum Building. Another significant formal element was the Rose Garden designed in 1916 by the eminent landscape architect Beatrix Farrand at the request of N. L. Britton and the board. The impressive triangular garden, inspired by the late nineteenth-century Roseraie de l'Haÿ outside Paris, was laid out, but none of the hardscape was created at the time because of a lack of funds. The Rose Garden was finally finished in the 1980s—seventy years later.

View of the library building and conservatory, 1921.

WPA workers building the rock garden designed by Thomas H. Everett, c. 1935.

In the early 1920s, the Garden commissioned the Olmsted Brothers firm to refine and expand the visitor circulation system. Their work is more or less in place today, although parts of the landscape have been eradicated by urban development—most significantly, a huge section on the east side that was chopped off in the 1930s by Robert Moses for the Bronx River Parkway. As far as we know, no further master planning was undertaken at NYBG until the 1980s, when the internationally known landscape architect Robert Zion was commissioned to create an entirely new plan, which was never realized. In the long interim after the Olmsteds, however, Thomas H. Everett, then head of horticulture, created the Rock Garden; the influential landscape architect Ellen Shipman—known for her flower gardens—was commissioned to create the Ladies' Border at the south end of the conservatory; and the renowned landscape architect Marian Cruger Coffin was brought in by then-trustee Henry Francis du Pont (her patron at Winterthur) to lay out the ornamental conifers and the lilac collection, both near the Rose Garden. And in the 1970s, Dan Kiley contributed some modest modernist designs for gardens in the conservatory precinct.

Early in our own era, a very simple master plan was designed for us by Geoffrey Roesch, a principal of the landscape architecture firm Coe Lee Robinson Roesch of Philadelphia. The most important design feature of the Roesch plan was the siting and layout of the Everett Children's Adventure Garden, near Azalea Way and not far from the conservatory. Later in my administration, we retained Polshek Partnership to think through visitor circulation, since we were now using trams, and the location of a new parking facility. Where in the world, in this National Historic Landmark, could you lay down surface parking for eight hundred cars? The answer, of course, was nowhere, and we were all frustrated but relieved—we needed the parking, but we really didn't want more cars inside the Garden.

During these highly charged years, we hired a roster of landscape architects and designers to restore existing gardens (Patrick Chassé in the Home Gardening Center, Lynden Miller for the perennial garden, Penelope Hobhouse and later Martha Stewart for the herb garden, Shavaun Towers to revamp the ornamental conifers) and to create new landscapes adjacent to new or restored buildings (Darrel Morrison for the native plant riverbank at the Stone Mill and Signe Nielsen for the rearrangement of the spruce collection in the Ross Conifer Arboretum near the conservatory). We also retained many imaginative designers to create temporary exhibitions in the conservatory—Susan Cohen, Madison Cox, George Sanchez, Patrick Blanc, and Raymond Jungles among others. Two celebrated English designers came to help us with charming projects—Rosemary Verey for an ornamental vegetable garden and the Dowager Marchioness of Salisbury for an "auricula theater"—a traditional showcase of Auricula primroses—in the herb garden. The landscape architecture firm Andropogon worked on the forest restoration. Sheila Brady of Oehme, van Sweden designed the new Native Plant Garden and the plantings in the Azalea Garden. We were awash with creative people.

But before some of these designers began, we needed a detailed master plan, so in 2006 we commissioned the internationally renowned landscape architect Laurie Olin to create one. It was an expensive project, but it resulted in an essential, detailed survey of infrastructure (a century of work below ground!) and a series of wise recommendations for siting new facilities—the Workers Operation Center, the Green Materials Recycling Center, and the Edible Academy—all new and ambitious facilities to be ranged up and down our eastern perimeter near the Bronx River Parkway.

OLIN, Laurie Olin's firm, also gave us a brilliant design known as the "Heart of the Garden." As we were drafting the strategic plan for 2009–15, we had recognized that the extensive set of landscapes comprising about seventy acres in the geographic center of the Garden needed renovation and probably redesign. The elements were the fifty-acre first-growth forest running north and south between Azalea Way and the Bronx River gorge, the hilltop azalea collection, and the old native plant garden just to the north of the forest. These three major sites were adjacent to one another, and all had been poorly maintained for a long time. In addition, the historic Lorillard Tobacco Company Snuff Mill on the Bronx River was sited just on the eastern edge of the forest. This large section of the Garden was very near the new Leon Levy Visitor Center and not far away from the Enid A. Haupt Conservatory. We recognized that it was a prime visitor attraction and determined to restore the whole complex of landscapes and the Stone Mill. OLIN, with Laurie Olin himself doing much of the creative work, recommended dramatically expanding the native plant garden and designed a unifying concept for the entire area, respecting the historic landscape but creating something very new. We put all four of these projects into the new strategic plan.

We began with the large and problematic azalea collection. Occupying the west side of Azalea Way, for about five hundred feet as the roadway descended from the visitor center to the bottom of Daffodil Hill, the azalea collection had been established in the 1920s and '30s underneath a high canopy of magnificent tulip trees, white oaks, and sweet gums. It had received virtually no horticultural care for at least one generation, perhaps two. With a member of the board who was hoping he could underwrite its restoration, we started exploring on foot. As there were no paths or pedestrian ways in, we bushwhacked through the weeds and broken glass and around the old, neglected azalea plants to explore the site. On its central hill, the highest point in the Garden, we discovered dramatic long views down into the forest and across the Bronx River to the western facade of the Stone Mill, which catches the waning sun on winter afternoons. Of course, we became excited about the possibilities, and we all felt that an expanded and rejuvenated azalea garden would be a marvelous attraction. So we retained Shavaun Towers, our collaborator on many landscape projects over a twenty-year period. Shavaun is especially sensitive to historic landscapes and gifted at designing curving pedestrian paths that are at once sympathetic with the layout of older pathways and also accessible. She laid out several walks, coming in from Azalea Way and from the historic crabapple collection at the bottom of the hill, and we began to think about the planting.

Since many species of *Rhododendron* are native to Asia, we decided to develop the azalea collection primarily with Asian specimens and use native Asian woodland plants for ground cover. This would complement the American species across the way in the soon-to-be-developed native plant garden. Sheila Brady, working with our horticulture team, designed the planting so that pinks, whites, and lavenders were on one face of the hill, and yellows and oranges were on the other. We were able to save many of the ancient plants original to the collection, and we then filled in with hundreds of new species and cultivars.

The Maureen K. Chilton Azalea Garden was opened in 2011. Peak flowering is in May, just around Mother's Day, so it has become a visitor favorite. But the landscape is beautiful year-round with the magnificent specimen trees, the ground-level planting, much of which is evergreen, and interspersed plantings of later-flowering shrubs such as hydrangeas. On a summer's day, when breezes cool the air on the hill, this high point in our landscape is a delicious place to be.

Just across Azalea Way is the venerable forest. Legend has it that a thousand years ago the Lenape occupied the Bronx River gorge and this forest. Some Native American materials, arrowheads, and other items have been found in the area and are now in the collections of the Bronx County Historical Society. It is also thought that the forest was a no-man's-land at the time of the Revolutionary War—the British on one side and established Dutch farmers on the other. The existence of this forest, the only old-growth forest in New York City, figured in the thinking of NYBG founders when they chose this site for the Garden. It had always been recognized as picturesque and precious. But by our era the forest had been overtaken by invasive species, some of them trees of considerable height and girth, and for a long time it had been abused by Garden visitors who had refused to stay on the paths, trampled the wild understory, and compacted the soil in many places. When the Garden was first established here 125 years ago, there had been a healthy carpet of native woodland flowers on the forest floor, but they were long gone—purloined by visitors a century ago.

To begin the restoration process, our horticultural staff, under the leadership of Todd Forrest, Arthur Ross Vice President for Horticulture, and his colleague Jessica Schuler, wrote a detailed forest management plan. Then, with the Andropogon firm, they set out to reestablish all the trails through the forest and monitor their use to prevent further compaction and to remove the most egregious invasive species. The worst was the Amur cork tree. This Asian native, a beautiful tree that had been in the horticultural collections of NYBG early in the twentieth century, had escaped into the forest, where it had aggressively established itself. It was everywhere, some trees twenty to thirty feet tall, and then there were millions of seedlings. Underwritten by an extremely generous gift from Carmen and John Thain, our horticulture staff undertook the massive task of removing all the Amur cork trees. It took about a year I think, executed without heavy equipment, so the trees could be removed in pieces and taken out on the mulched trails.

Then we began an extensive program to reestablish native species in the forest, both trees and shrubs, and our forest team began to collect acorns to grow seedlings for reforestation, clones of the existing trees. In the past this forest had been known as the "hemlock forest in the Bronx" because there were thousands of native hemlocks here. When I came to the Garden there were only a few hundred left, the rest destroyed by the wooly adelgid. Today only a handful of the original hemlocks remain. The other species, in better condition today, are white oaks, tulip trees, ash, some maples, and some sweetgums. The site is not highly biodiverse, although our team is still planting appropriate species among the older trees and shrubs. The Thain Family Forest restoration was complete and opened to the public in 2011.

The old native plant garden, established almost a hundred years ago to promote the understanding and conservation of the plants of the northeastern United States, and a favorite undertaking of Elizabeth Britton, was just to the northwest of the forest and very close to the new Leon Levy Visitor Center. The original garden was so modest and so underwhelming that we decided to create a new garden. In an informal competition, we invited several landscape architects to submit designs, using a plant list we provided, for a large-scale, three-acre garden with a water feature in a contemporary style. We thought it was time to show our public what a contemporary garden would look like in our historic landscape. Sheila Brady's beautiful concept won.

Nearly one hundred thousand native trees, shrubs, wildflowers, ferns, and grasses surround three stepped and serene pools, the water flowing downward over two weirs. Technologically contemporary as well, the feature is designed to collect rainwater from the site and recirculate it, sparing the city water mains we had previously used for Garden water features. Hugh Hardy designed two small pavilions, one at the entrance for groups to gather under and another for school group programs. He and Sheila decided that the visual vocabulary for these shelters and the surrounding stone walls and fences should be the same that Hugh had used at the nearby visitor center. Sited along Azalea Way, a main thoroughfare from the visitor center past the forest to the Peggy Rockefeller Rose Garden, the new Native Plant Garden opened in May 2013. Its contempo-

View of the new Native Plant Garden. Photo © Marlon Co.

rary design is fresh and surprising, but it blends seamlessly into the historic fabric of the surrounding landscape.

Immediately across the river to the east of the forest is the Horticulture Operations Center. This behind-the-scenes complex, created to serve the needs of our sixty-plus gardeners and arborists, took many years to complete, largely because fundraising for such a functional project proved difficult in the extreme. There was some city money available, but the project required about $10 million in private funding, and there were not many donors interested in the renovation and expansion of horticulture support facilities. It was necessary to do a gut rehab of an old headhouse to create new locker rooms, a lunchroom, safe pesticide storage, and other support facilities. In addition, there were several old greenhouses, many torn down, several restored, on the site. It was also necessary to create a massive new facility for the storage of compost and potting soil (some years, between eighty and ninety thousand plants are potted up at NYBG using potting soil we create), and we undertook the restoration of a series of charming pit houses for the propagation and housing of thousands of plants. And we needed a structure for the Garden's scores of mowers, blowers, golf carts, and trucks used on a daily basis by the horticulture team. This complex of functional buildings surrounds two open spaces, one a parking lot for the staff who use this as their headquarters, and one a large quadrangle used as a nursery for trees and shrubs.

Also in the Heart of the Garden, on the eastern bank of the Bronx River adjacent to the forest, stands the Stone Mill, a National Historic Landmark and a New York City Landmark. Built by the Lorillard Tobacco Company from stone quarried on the site, the mill was used for grinding tobacco into snuff. The Lorillards had owned about a hundred acres along the river since the 1790s and had built several wooden mills, but replaced them around 1840 with this fine stone structure. The family and business moved away in 1870, and the land was acquired by New York City in 1884, which transferred it to NYBG in 1915. The mill was used as a storage facility by the city parks department or stood empty for several generations until it was rehabilitated by NYBG in the mid-twentieth century as a restaurant and entertaining facility. By the 1990s that earlier renovation needed total rethinking, and we undertook a complete

The restored Goldman Stone Mill. Photo © Robert Benson Photography.

overhaul, installing a system of stormwater management for runoff into the Bronx River, designing historic interiors for entertaining, and creating a suite of offices for our horticulture staff. Complemented by an extensive planting of native trees and shrubs designed by Darrel Morrison on the riverbank hillside, the refurbished Lillian and Amy Goldman Stone Mill, funded by the City of New York and Amy Goldman, our devoted trustee, opened with considerable fanfare in 2010.

We turned our attention next to two large areas not far from the Heart of the Garden, which feature woody plant collections that are both taxonomically significant and beautiful. The Burn Family Lilac Collection, with hundreds of venerable and newly established plants, was originally designed by Marian Coffin in the 1950s. Our generous trustee Jeanie Burn had been inspired by the lilacs at Harvard's Arnold Arboretum and funded the restoration of our collection so it would be equally beautiful. Shavaun Towers designed the new garden, which opened on a glorious May day in 2016. Just across the way, on a stony hilltop overlooking the Rose Garden, is the new Judy and Michael Steinhardt Maple Collection. Our trustee Michael Steinhardt, a well-known financier who has one of the world's finest

The Marjorie G. Rosen Seasonal Walk by Piet Oudolf. Photo © and courtesy Rob Cardillo.

collections of Japanese maples in his home arboretum, helped us reorganize, restore, and enlarge our historic collection in an elegant landscape also designed by Shavaun Towers, which showcases both mature trees and new plantings.

Across the Garden, adjacent to the conservatory, a brand-new garden was created for us during these years—the Marjorie G. Rosen Seasonal Walk by the Dutch designer Piet Oudolf, a leader of the "New Perennial Movement." Marjorie is a long-term trustee, and she wanted us to be showing the very fresh design theories of this contemporary landscape artist. The two-hundred-foot-long double perennial border adjacent to the conservatory is quintessential Piet, featuring tall prairie plants and grasses and many new cultivars of old garden favorites. It's at its most exciting in the late summer and fall.

During this same period of intense activity, we undertook something very different—a huge building nowhere near the Heart of the Garden, but close to my heart. It was for us a staggering capital project: the construction of a parking facility adjacent to the Botanical Garden station of the Metro-North Railroad. The charming Victorian station building has been in place for about a century, since the Botanical Garden stop was established by the New York Central Railroad in the 1890s by Cornelius Vanderbilt II, then the president of the railroad company and president of the board of NYBG. It was a lifeline for the Garden. But times changed, and more people drove to places they wanted to go. As attendance grew through the 1990s, and weekends became very popular with Garden visitors, there was never enough parking. No large-scale surface lots had been built at the Garden in the 1950s as they had at the Bronx Zoo and other destinations. Over the years various master plans had attempted to locate an extensive new parking facility somewhere inside the property, but the Garden's rolling landscape, with its magnificent old trees and rock outcrops, provided no appropriate site. I was very glad that this was the case, as I certainly didn't want to be the builder of asphalt lots. But we were in a tight spot as we were even beginning to turn visitors away on popular days because there was no place for them to park.

Then a remarkable thing happened. One day when Bill Steere Jr. and I were talking with George Pataki about other Garden needs, the governor suddenly asked, “Is there any private land for sale around the perimeter of the Garden?” Bill and I looked at each other with surprise, and I explained that there really wasn’t much private land at all around the perimeter because the Bronx Zoo, Fordham University, and the Bronx River Parkway occupied three sides. “What about the north?” he asked. “Well,” I said, “that’s the Metro-North tracks, and just across the tracks there are some old industrial buildings which perhaps we could buy if we had a use for them.” Pataki asked us if $10 million might buy one of those sites, and we told him that it was very likely. The upshot was that the State of New York gave us the funds to buy one acre on the Metro-North tracks adjacent to the train station, on our neighborhood street Webster Avenue, and that’s where we decided to build our long-desired parking structure.

But, of course, the parcel we wanted, home for most of the past century to a family-owned boiler plant, was not for sale. For several years we tried to acquire it, but it was only when I turned over the task to J.V. Cossaboom, by then NYBG’s chief operating officer, that we succeeded. J.V. had become like a brother to me and was my closest confidant and counsellor. He can be pretty persuasive, but it even took him about eighteen months of long lunches with the owner to close the deal. Finally, though, we were able to tear down the old building, remediate the polluted site, and begin the project. The Bloomberg administration and the city capital budget for cultural institutions were central funders, and we also raised private funds. The $39 million parking garage was designed by Susan Rodriguez of Ennead Architects as a green building, covered with vines that rise to the top of the five-story structure.

Around this same time, another large parcel of land on the corner opposite the garage became available, and we decided to purchase it (without state funding this time) as a likely place for the development of affordable housing. And as of this writing, a good developer is planning a 449-unit apartment building. It is now fully government funded, and as we own the land lease, the project will yield a handsome return to NYBG.

The business plans for these undertakings, devised by J.V., were a new kind of venture for us. The garage is used by Garden visitors on weekends, when we really need it, and by commuters and members of the community on weekdays. It's a tremendous source of revenue because it's 100 percent full all the time. The perfect solution to one of our thornier problems. Not as beautiful or as much fun as the Native Plant Garden, but perhaps just as important. And the apartment building, itself a fine improvement to the old Webster Avenue streetscape, will be a very worthwhile investment.

In retrospect, this third strategic plan seems rather monumental, and I'm thinking about how we actually accomplished it all. First, the plan was well thought out and stretched over seven years, although as usual one or two capital projects took longer. And we had a board and a fundraising apparatus that could generate the $480 million. And we had the indomitable J.V. overseeing a crackerjack capital projects office managed by our in-house landscape architect, Ursula Hoskins. And we had the Bloomberg administration at our back. And we were all fairly young and wildly determined. In the end, I hope it will be inspiring to the leaders of other organizations to see that an ambitious and comprehensive strategic plan created bottom-up by staff and trustees can provide the structure and impetus for this kind of institutional weightlifting.

Our second major focus in this period was development of special exhibitions and marketing. These days, special exhibitions drive audience development and earned income for many museums. But NYBG's special exhibition program was very modest. "Very modest," until 2007 when an ambitious new era was launched with *Kiku: The Art of the Japanese Chrysanthemum.* Our senior horticulture figure at that time, Kim Tripp, had made special arrangements with the "kiku master" at the Imperial Household in Tokyo, and over five years of apprenticeship there our horticulturalists were taught the ancient art of extravagantly trained chrysanthemums. Accompanied by a gallery exhibition of Japanese garden art, the show was a great hit and has been revived every year since. Following *Kiku*, in 2008 we mounted *Darwin's Garden: An Evolutionary Adventure,* curated by David Kohn, a noted Darwin scholar, in the library gallery. This was the first of what has now become an extensive series of multimedia Garden-wide exhibitions illuminating the interrelationship of nature and culture, science and art. *Darwin's Garden* featured rare books, manuscripts, and letters documenting Darwin's thinking about plants and was accompanied by an installation in the conservatory that re-created Darwin's garden in the English countryside. Mixed herbaceous borders overflowed with foxgloves, delphiniums, larkspurs, roses, silver artemisias, and climbing clematis. Darwin was an avid gardener, and his garden was the scene of some of his experiments—the breeding of primroses and his study of orchids, for example.

This exhibition launched an extensive series of similar shows about cultural figures and the gardens that inspired their work. Since *Darwin,* we have mounted *Emily Dickinson's Garden: The Poetry of Flowers* (2010), *Monet's Garden* (2012), *Frida Kahlo: Art,*

The "Thousand Bloom" ozukuri on display in *Kiku*.

Garden, Life (2015), *Georgia O'Keeffe: Visions of Hawai'i* (2018), and *Brazilian Modern: The Living Art of Roberto Burle Marx* (2019). Other major special exhibitions have focused on places or works of art, such as *Moore in America* (2008), *Spanish Paradise: Gardens of the Alhambra* (2011), *The Italian Renaissance Garden* (2013), *Groundbreakers: Great American Gardens and the Women Who Designed Them* (2014), *Impressionism: American Gardens on Canvas* (2016), and *CHIHULY*, the work of the world-renowned glass sculptor (2017). Until she left NYBG in 2017, the accompanying flower shows were created by Francisca Coelho with the fantastic support of plantsman Marc Hachadourian and his crew in the Nolen Greenhouses.

In my era, the special exhibition program was supported by dedicated staff reporting variously to four senior executives: the CEO, the vice president for horticulture, the director of the LuEsther T. Mertz Library, and the director of the Garden. In addition to fundraising and marketing staff, others dedicated to exhibition development and execution include horticulturists in the Enid A. Haupt Conservatory and Nolen Greenhouses, exhibition planners, a conservator, a registrar, and two interpretation specialists who produce

Monet's Garden **exhibition in the conservatory, 2012.**

Frida Kahlo: Art, Garden, Life in the conservatory, 2015. Photo © Ben Hider Images.

The garden created by Raymond Jungles for *Brazilian Modern: The Living Art of Roberto Burle Marx*, 2019. Photo © Marlon Co.

all the written materials including signage and an exhibition catalog. In conjunction, there are poster exhibitions in the Mertz Library Rotunda and in the Ross Gallery; and lectures, symposia, and performing arts events. We also always develop a "poetry walk"—an outdoor display of poems related closely to the theme of the exhibition (a joint project for years with the poetry scholar Alice Quinn and the Poetry Society of America).

The process of developing ideas for exhibitions and devising a full program is not linear—it is highly creative, even intuitive, and ideas come in various ways, from staff, acquaintances, reading, travel experiences. All ideas must lend themselves to our multidimensional, multimedia format and are tested by our marketing department in e-surveys and focus groups. When an exhibition concept doesn't pass muster in testing, we either recast it or we drop it. It customarily takes 24 to 36 months to plan and realize an exhibition, and as we feel the necessity to mount one every year, we are often working on three simultaneously. When an idea has proved market-worthy, we locate a subject-specialist curator, a scholar in the field, who can be retained as an independent contractor. For example, our research turned up the marvelous Adriana Zavala, an art historian from Tufts University specializing in twentieth-century women artists in Mexico, to create our Frida Kahlo show. The concept is developed, and we decide which art to borrow, relying on the credentials of the curator—otherwise why would museums in Mexico City lend 70 million dollars' worth of Frida Kahlo's art to a botanical garden in New York City? Simultaneously, the companion conservatory exhibition is designed, and in the fall of the year before the show is to open, we make the plant list, order the seeds, and begin to grow the plants. The catalog is written and the surrounding events are created. The exhibitions usually run three to five months, spring to fall. And while all of this is going on, our creative team must also mount an orchid exhibition with a different theme every year, a *kiku* display, and the annual *Holiday Train Show*.

Exhibitions combining horticulture, landscape, garden design, and the visual arts draw huge audiences. As a result of these exhibitions, Garden attendance grew from approximately 700,000 in 2008 to 1.3 million in 2017. Pre-pandemic, we credited the special exhibition program with generating 30 to 40 percent of our annual revenue, grossing $25 to $30 million per year from ticket sales, group tours, parking charges, food service, retail, membership upgrades, and underwriting grants. Our special exhibitions have often received major support from the National Endowment for the Humanities, who can see that they reach new humanities audiences—our venue is unique among American museums. In addition, our exhibitions have earned lavish praise from the media, including National Public Radio, the *New York Times,* and the *Financial Times* of London. They have been central in establishing our reputation as a world-famous place of intellectual excitement and beauty.

Sigourney Weaver came into the orbit of the New York Botanical Garden a decade ago when we invited her to be a special guest at our annual June Conservatory Ball, our principal social event of the year. Sigourney is a dyed-in-the-wool New Yorker. She was born in New York City and has lived most of her life on Long Island and in Manhattan. Hollywood royalty she may be, but Sigourney loves New York. She is also known worldwide for her passionate environmental activism. Sigourney sat next to me at dinner that evening, and I regaled her with the glories of NYBG, especially our role in scientific research. I invited her for a golf cart tour of the Garden and all our facilities, and we spent the day together a few weeks later. Since that time, her connection to the Garden has deepened, and she joined the board of trustees in 2011.

Sigourney's contributions to the Garden have been profound. She has been the voice of NYBG, promoting our programs in the media, speaking at our events, hosting the PBS *Treasures of New York* series episode on NYBG, and lending her celebrity to boosting our reputation. And she gave us an especially memorable gift for our *Georgia O'Keeffe: Visions of Hawai'i* exhibition. In 1939, O'Keeffe had sailed to Hawai'i to paint tropical plants. Most of those paintings are in the collections of the Honolulu Museum of Art and had

On a Garden tour, Larry Condon, Sigourney Weaver, Gregory Long, and NYBG friend Helen Mirren. Photo © Marlon Co.

not been seen in New York since before World War II, until our exhibition. But we managed to borrow the entire set (or all but one, that is) for a gallery show, and we created a lush Hawai'ian garden in the conservatory and made a ten-minute film in which Sigourney reads O'Keeffe's emotional letters to her estranged husband, Alfred Stieglitz, who was back in New York. In Sigourney's sonorous voice, the letters had an ineffable, dreamy quality, and hearing them was a memorable experience for hundreds of thousands of visitors to the exhibition.

In spite of her international travel schedule, Sigourney tries every year to see our major exhibitions, and she is usually at the Conservatory Ball. She has been a stalwart friend, and now many trustees know her and love seeing her when she's visiting, and many of our social friends know she will probably be at the next party. One year, when she was in Los Angeles in the middle of making a new *Avatar* film, she arranged with James Cameron to take a few days off to fly to New York to appear at the Conservatory Ball. The party was in my honor to mark my retirement, and, as it was during our *Georgia*

O'Keeffe exhibition, she made a charming presentation at dinner, part of it in Hawai'ian—without notes!

Although we often have displayed works of art in conjunction with temporary exhibitions, we have decided that NYBG would not install permanent artworks in the Garden. Our temporary exhibitions have included outdoor installations of sculpture by Henry Moore and Manolo Valdés, Dale Chihuly's blown glasswork, and tile mosaics by Roberto Burle Marx, and inside we have shown paintings in our gallery. But we have resisted the frequent offers of gifts from artists and their families, dealers, and collectors that come with the proviso that the work will be exhibited in our glasshouses or outdoors. For one reason, we simply do not have the expertise to acquire, conserve, or interpret art. Our temporary exhibitions have been curated by outside experts who are contracted independently. Their expertise is an essential component of our successful special exhibitions program. If we were to install art in the Garden on a permanent basis, we would have to have art professionals on staff, and we have never felt this to be affordable, advisable, or necessary.

But more important, we strongly believe that the permanent collections of NYBG are the plants. We are collecting, exhibiting, and interpreting our one million living plants for purposes of public education. The installation of works of art, we feel, would diminish focus on the plant collections and their value. Our special exhibitions, often with art as a major component, come and go. Our landscapes (designed and natural) have remained uncluttered and pristine. They are foreground, not background.

NINE

My Final Years and the Edible Academy

In the spring of 2018, just one month before I retired, we opened the Edible Academy at NYBG, the last of the twenty-plus capital projects we had accomplished in our era. Comprising a barnlike interactive classroom building with a teaching kitchen, a freestanding greenhouse and teaching pavilions, an amphitheater, composting toilets, and large garden plots, this project for teaching children how to grow organic vegetables, cook what they grow, and live healthy lives had taken six years to conceive, fund, and build. It is the culmination of a seventy-five-year history of teaching New York children vegetable gardening.

Since the nineteenth century botanical gardens have been devoted to the idea of horticultural education for children, although for many generations the focus was primarily on flowers. At NYBG, the first garden established to teach children about vegetables was created in 1950. Ever since, NYBG has maintained a dedicated space for children's gardening education, and the Edible Academy is the newest and most fully developed such facility. The idea for it stemmed from our belief in the exceptional talents and energy of the gardener-teachers who had been running our modest seasonal Family Garden program for several years. Simply said, we felt emboldened to build this large-scale project because of them. And our belief was well founded: ten years since we had the idea and now three years since it opened to the public, they're all still there.

With the burgeoning farm-to-table movement, and the tremendous interest from schools in gardening for children, the audience for our vegetable gardening program, which operated in most years from April to early October, had grown to fifty thousand a year by 2010. Our program had to be seasonal, as our classroom was just one large garden plot. And it was pretty rudimentary—there were no restrooms for school groups and nowhere to conduct classes in the rain except under two makeshift picnic shelters. The Family Garden's simplicity gave it great charm, but the drawback was that classes couldn't begin before spring and had to close down before cold weather in the fall—we were missing most of the school year. We realized that this seasonality was a problem when the children, their parents, and their teachers began to ask for year-round programming. For a few years, we held classes through the fall in big tents accommodating large weekend audiences, but it was too cold to continue from November to March.

Meanwhile, at the same time that teachers were talking to us about year-round gardening content, the school garden movement was spreading. In New York City, school officials wanted to replace their parking lots with gardens. But as schools were closed from June until September, and in the Bronx many teachers were commuting from the suburbs, the only school staff available to maintain a garden through the summer were custodians, and often they were not prepared or trained to maintain vegetable gardens through three difficult months of the year. There was also the problem, which all vegetable gardeners will recognize, that if zucchini plants are happy and healthy, there are too many zucchini—and no teachers or children on-site to eat them!

With all this in mind, the government relations staff at NYBG went with our educators to the city council and asked for major support for a year-round vegetable gardening program at NYBG—a school garden for all the Bronx. The city council leadership, many of whom were Bronx elected officials, loved the idea. And so, with initial funding promised by New York City, we set about designing the new Edible Academy.

We knew we would need for the winter months at least two

indoor classrooms for school groups Monday through Friday and drop-in visitors, mainly families, on the weekends. We knew we would need a greenhouse for propagating tens of thousands of vegetable seedlings every spring and for growing crops under glass in the winter. We knew we would need a barn for storing equipment and tools. We knew we would need shelters for school groups on sunny and rainy days. We knew we would need new restrooms to accommodate crowds on the weekend and school groups during the week. We knew we would need offices for staff and volunteers. And we knew we wanted a "green" project.

Our first step was to think about the scale and design of the buildings and layout of the site. The half-acre Family Garden had lovability, and we certainly wanted to maintain it and not to diminish its special qualities. Positioned on a high site, with no overhanging trees, the garden has broad exposure from the east, south, and west all day and long views downhill to the south and to the west, into the deep rock-cut gorge of the Bronx River. It's set on the footprint of the former Lorillard mansion, which had been taken down long ago, and we wanted to respect the historic site, so we decided on a cluster of small buildings at the northern end of the Family Garden, which would feel agricultural but with a contemporary look.

We began to consider architectural firms who could design such a complex—an indoor/outdoor school, various sheds and pavilions, a greenhouse, a plant nursery, large-scale restrooms, and offices, surrounding the featured Family Garden, which would remain on its original site in the center. Over the years we had done a few projects with the New York architects Cooper Robertson, and we chose them because they had expertise in school buildings and museum planning and design. In their analysis of the site they discovered that to the north and downhill from the central garden there was a perfect bowl that could easily be transformed into an amphitheater for about 250 people—a stroke of topographical luck.

The Edible Academy program is broader than organic vegetable gardening. There are classes in cooking, beekeeping, and healthy lifestyle practices, and the facility itself teaches many aspects of sustainability and green building design—geothermal heating and

View of the Edible Academy, 2018. Photo © Robert Benson Photography.

cooling; solar collectors that provide all the electricity needed to light the buildings; a green roof on the classroom building planted with succulents and herbs; stormwater capture; and composting toilets (which have been a big success). There is state-of-the-art composting—no inorganic fertilizers are used. The Edible Academy produces so much food that nearly everyone who comes in summer and fall can take home something to eat.

In its first full year of operation, the Edible Academy hosted one hundred thousand people, double the previous year's (seasonal) attendance. The facility was designed to serve one hundred thousand, but we had no idea that that many would come in the first twelve months. In that opening year, the first three weekends were attended by a total of twelve thousand people—weeding, harvesting, attending cooking demonstrations and beekeeping sessions. There was plenty of music and dancing as well!

Over the years, as our ambitions for the project had grown, so had the capital budget, which ended up at $28 million. The money was raised about fifty-fifty from public and private sources. In addition to the city council, the state helped, but, surprising to me, the private side of the fundraising for what I thought was a heartwarming project was difficult and took more than three years. It was coming at the end of our third big campaign, and we had already raised nearly a billion dollars cumulatively, so perhaps everyone was just tired. But in the end, we did reach our goal, and the project was fully funded. We were even able to generate some new endowments—the income from which was restricted to Edible Academy operations.

During the tough fundraising, I often wondered whether this large expense would be justified. But now that the Edible Academy is teaching a hundred thousand people a year about organic gardening and healthy diet, I'm happy we took the leap. The Edible Academy will serve the people of the Bronx and New York City area for many generations to come.

Intended to be a centerpiece of our third strategic plan, the Edible Academy was way behind schedule. We had hoped it would open by 2015, but the slow fundraising delayed it until 2018. However, all of the other principal elements in our third plan had been accomplished, and so we began to look forward once again. As we were on the cusp of our 125th anniversary, we felt we needed a new logo. "NYBG / 125" was created for us by the graphics firm Pentagram, and we named our fourth planning document the *125th Anniversary Strategic Plan.*

At this juncture we decided to create a six-year program, for two reasons: first, we had already done three seven-year plans and we were a little tired of that format, and second, we realized that when we got to the sixth year of this plan, in 2021, I would be seventy-five—too old? So, not only did we set a limit of six years, but we also divided the plan into two three-year phases: 2016–2018 and 2019–2021. Our overall goal for the first phase was $125 million, and we had generated that before I left. Our six-year goal for new endowment was also $125 million, but no consolidated goal was set for the second three-year phase. We left that open.

Of course, it was in the back of my mind, and perhaps the minds of some of my colleagues, that I would oversee the initiatives we laid out in the first phase, but that I might hand over the second phase to my successor. And, as it turned out, I did complete Phase I of the plan by the time I retired, and I hoped the second phase would provide a leg up for my successor. Reflecting the many changes we had made in the shape and functioning of the organization but also sticking close to our institutional mission, the new six-year plan was conceived in thematic sections: "Creating a Green Urban Oasis," "Connecting Gardening to the Arts and Humanities," "Teaching Science to City Kids," "Saving the Plants of the World," "Anchoring the Community," "Business Activities," and "Management and Institutional Needs."

In 2015 we were a mature organization with well-defined commitments to established programs and well-honed, successful ways of operating, so this fourth plan was more about consolidation than innovation. We were interested in continuity. We were able to draw on our knowledge of what our constituents cared about. We had come to a deep understanding of our audience. And so we had fun devising new ways in the plan of talking about our traditional interests: for example, we did not title the science section of the plan "Plant Research and Conservation," but instead we decided on the more compelling "Saving the Plants of the World." In addition to the development of horticulture and living collections, we had become very focused on audience development, visitor services, and public education, and so this section was titled "Creating a Green Urban Oasis." Our exciting programming in exhibitions and related activities we called "Connecting Gardening to the Arts and Humanities."

For the "Green Urban Oasis" section, we identified two major capital construction projects. First, we needed to repair the compression ring in the conservatory dome—as it turned out, quite an expensive job when combined with necessary energy upgrades including new boilers and backup generators. And second, the conservatory had turned out to be a less-than-ideal venue for some of our bigger special exhibitions, so we decided to plan a new exhibition house, which, when it was built (later, we assumed), would be the site of some of our major shows, including the *Holiday Train Show*. Many landscape improvements and restorations were also envisioned—a huge new planting of daffodils to expand our historic collection, a project we named "A Million Daffodils!"; improvement of the Daylily/Daffodil Walk; restoration of the as-yet-unimproved first-growth woodland south of the Thain Family Forest; rejuvenation of the Rock Garden; and complete restoration and expansion of our Magnolia Collection.

In "Connecting Gardening to the Arts and Humanities," inspired by our past successes, we devised a new series of special exhibitions: these included the Garden-wide installation of glass sculpture by Dale Chihuly (2017); the creation of the Hawai'ian garden in conjunction with *Georgia O'Keeffe: Visions of Hawai'i* (2018); and *Brazilian Modern: The Living Art of Roberto Burle Marx,* a celebration of the innovative artist, landscape architect, and conservationist (2019). Looking ahead, we invited the revolutionary Japanese artist Yayoi Kusama to create a one-woman extravaganza for 2020 (which was postponed to 2021). In this section of the plan we also included our hope to extend the Humanities Institute in the Mertz Library and raising funds for the library archives.

In "Teaching Science to City Kids," we focused on the renewal of the Everett Children's Adventure Garden, much-needed after almost two decades of hard use by as many as 250,000 people every year. We hoped to accomplish classroom renovations in other buildings, and we moved forward with our nearly finished Edible Academy. In the section "Saving the Plants of the World," we committed ourselves to generating funds to support our ongoing expenses in our established scientific programs. These range from ecological studies in New York City, to basic plant research in the Amazon, to field biology supporting conservation in the coastal forests of Brazil and biocultural diversity in the South Pacific. "Anchoring the Community" involved opening a new entrance, now known as the East Gate, to provide access to the Garden by pedestrians coming in from the neighborhoods to the east. We also thought through new outreach programs for changing audiences in the Bronx and renewed our commitment to our long-standing Bronx Green-Up program.

In all of our strategic plans, we never ignored financial or personnel needs in business activities—especially marketing, communications, and customer relations—or general needs in institutional

The Honorable Ritchie Torres, congressman from the Bronx (center), with NYBG trustee Joseph Thompson and Gregory Long at the dedication of the refurbished East Gate, 2017. Photo © Marlon Co.

management, and once again these were identified and prioritized. The financial plan called for the establishment of a steady state and balanced budget by the year 2022—when all had been accomplished. But of course, along the way, every year as we raised the funds and realized our goals we also balanced the operating budget. And since we were celebrating the institution's 125 years, it was natural to settle on the concept of creating a new endowment fund of $125 million. A substantial portion of that new endowment has been raised, and at the end of December 2020, the institution's total endowment was approximately $373 million.

In 2016, when Frida Kahlo fans and our female mariachi band were carrying on at NYBG, and I had worked at the Garden for twenty-seven years and was approaching my seventieth birthday, I suffered a year of intense ambivalence. I loved the Garden and was hugely proud of everything we had accomplished. It is not too strong a notion to say I loved the most active and committed trustees and my senior staff—they were like family to me. We were all working together in a carefully choreographed and strategic manner, and we were all, I think it is fair to say, at the top of our game. The institution had never in its long history been in better shape. But in spite of all of that, I had begun to feel that my strong-willed leadership of the institution over three decades might become, in the long run, a liability of sorts. I felt that there was too much emphasis generally on me and my thinking. I had become a kind of benevolent dictator, if you will. I struggled with the tension between wanting to extend my happy years at the Garden and sensing that it had probably been enough—for the institution especially, but probably also for me. So, in April 2017, I announced to the board, in a speech over which I had labored mightily, that I would be stepping down from the presidency in June 2018. I was giving fourteen months' notice—a lot, I thought, but not too much.

We began to think about the transition, which was a little tricky, and in this I depended greatly on the experience and fine judgment of J.V. Cossaboom, who had assumed the "directorship" of NYBG a few years earlier. Our first task was to address board leadership. The chair of the board had served two three-year terms and two years of her third term and was planning to step down in fall 2018. No chair had ever served for more than three three-year terms, and precedent required that she retire. But that meant that the two of us would be leaving the Garden in the same year and within about six months of each other. The institution was very strong at this point and probably could have weathered two new figures at the helm, but with a small group of trustees, J.V. and I decided to create a transition plan that would make these changes in management smoother and easier to absorb.

With the chair of the nominating committee, Caroline Wamsler, we developed something we jokingly called RLP—"Responsible Leadership Planning." We decided that a new board chair needed to be identified and named "chair designate" sooner rather than later, considerably in advance of my retirement and of the formal election of a new chair in November 2018. In this way, the new chair would be known before my successor assumed office, and it would be a well-known figure, a respected board member, someone whose presence in the top board slot would be reassuring during that time of CEO transition.

However, our best-laid plans didn't quite work out. The board did name a chair designate, and for a while all was well. But after a few months she decided that her other professional commitments would make it impossible for her to accept the position. We were distressed, of course, because she would have been perfect, but fortunately another highly respected board member was willing to be named chair. He was elected chair designate by the board in January 2018. And so, by the time I retired on June 30, 2018, everyone felt comfortable with the new chair; my successor as CEO had been hired and was attending briefing meetings; and we had also, at my last meeting, elected two promising new board members.

Meanwhile, during my last six months, I was also occupied with preparing the Garden in programmatic and financial areas. I felt we had done everything we could do to reassure the board, the staff, and our donors and members that the leadership transition would be smooth and elegant, but there were many mid- and several long-term program plans I wanted to put in place to hand over to the new administration. So we continued to work on the initiatives detailed in the strategic plan. But then some new work appeared—as it always does.

On May 15, 2018, during the reception for our members and patrons to preview the *Georgia O'Keeffe* show, a violent thunderstorm appeared, and we watched from the library as a giant tulip tree at the top of our aging double allée was struck by lightning and wrenched out of the ground, toppling down across the drive and lawn. We knew that this magnificent set of trees, planted more than a century ago, was senescent—they had been dropping large branches for several years—and we had recognized that they would have to be replaced soon. A few months earlier, in fact, I had asked Laurie Olin, our master-planning landscape architect friend, and our new trustee Robert Bartlett, of the long-established Bartlett Tree Experts, for help. In collaboration with several other trustees and the CEO designate, we had created a plan for their replacement, hoping of course that it could be put in a drawer and only be brought out much later when the dreaded day for replacement would arrive. But, because of the May 2018 storm, one tree was already down and, earlier than any of us had wished, we had to begin the process. But fortunately we were ready. Our first step was to buy a set of twenty-six matching trees so that when the replanting was accomplished, which we knew would take three or four years, the trees would all be the same size. Now, three years later, more old trees have failed and are down, and the Olin plan is being implemented. The original formation, planted in 1903, had lived intact until 2018. By 2024, a new allée will delight visitors—hopefully, for at least another 120 years.

During this time, we also focused on our financial future. J.V. and I made sure that our revenue picture for 2018 would result in our twenty-ninth consecutive balanced budget. Also, after much careful research, we concluded, with the finance committee of the board, our months-long process of handing over the management of our endowment to a new consulting firm, and I am delighted to say that they are doing sterling work for us.

New board and executive leadership had been chosen, a great deal of necessary programmatic planning had been accomplished, major exhibitions were in development for the next two years, a balanced budget in the current year was guaranteed, and solid endowment management was in place. In addition, the trustees and I had created a new "President's Fund," a $39 million fundraising success of both multiyear operating and endowment gifts made in my honor. And, although it was not large, we had managed to salt away a nice rainy-day fund of about $3 million as well. In most respects, our carefully constructed Responsible Leadership Plan had guided us successfully to the end of my administration. We were ready to hand the institution over to a new leader.

I believe that in the lives of organizations, ceremonial events are very important. They emphasize the significance of the organization. Plus, they can be great fun since a party is often involved, and they provide the perfect setting for bonding among trustees and staff. Like many museum directors, I am a producer at heart, and I have always loved conceiving and putting on these celebratory events. As my retirement date approached, the institution was in very fine shape, and I felt we had much to celebrate. So, we set the date for a going-away party—June 20, 2018. And what a festivity we created!

That afternoon trustees, staff, patrons, friends from government agencies, corporations, and private foundations, neighbors, and close friends gathered in the Arthur and Janet Ross Lecture Hall. Eight speakers—Maureen Chilton, our outgoing chair; Carrie Rebora Barratt, CEO designate; J.V. Cossaboom; Dan Hinkley, plantsman and cofounder of Heronswood Nursery and Garden; Gloria Coruzzi, Carroll and Milton Petrie Professor of Biology at New York University; Bronx borough president Ruben Diaz Jr.; our vice chair Shelby White, speaking for Vartan Gregorian, who was not able to attend; and Larry Condon, senior vice chair—and I made short presentations, accompanied by beautiful images on the screen behind us. We spoke on two themes: what we had been up to in the past twenty-nine years, and our vision for the future.

There were 425 of us in that hall, more than capacity, and that meant that there were 425 intensely proud people who had worked hard—some for decades—to create the successful institution we were handing over to our successors. I felt it was important to set up my successor with an exciting and optimistic picture of the future of the institution, in which I most heartily believed, and to talk about the work of the organization and its role in the world rather than just all we had done. This extravaganza was extremely moving for me, of course, but also for the hundreds of friends and colleagues present, all of whom had participated in our institutional transformation, all of whom were invested in it, and all of whom loved the Garden. At the end of the program, Maureen Chilton made a surprise announcement—the board had voted to make me President Emeritus of NYBG. I was hugely honored.

We filed out, rather slowly, with lots of congratulations all around and thanks to the very distinguished speakers, each of whom had spoken from the heart and each of whom had their own quite dif-

Gregory Long wearing the NYBG Gold Medal, 2018.

ferent perspective on the institution and its future. We walked as a group to a big cocktail reception at our restaurant, the Hudson Garden Grill, and since it was a beautiful afternoon, the bar and drinks were set up outside on a terrace overlooking the conservatory. This was followed by dinner for 250 in the Garden Terrace Room. There the board had another surprise for me—the Gold Medal of the New York Botanical Garden.

Over our years together, we had staged many milestone events, all of which had built our camaraderie and esprit de corps. But there had never been one quite like this. Clearly it was a special moment for me, but more than that, I believe it held great meaning for the institution and the people who attended. Of course, we were all proud of our decades of work together, but more significantly, I think we all recognized that we were paving the way—to a fine future for the place we had been nurturing since 1989.

The reflecting pool in the conservatory's Palms of the World Gallery. Photo © Marlon Co.

APPENDIX

Our Planning Process

I would like my readers to understand our strategic planning methodology. As we devised it in 1990, it is a bit complex. In that year, in the first of our planning efforts, we hired a highly competent in-house facilitator, whose salary was funded by a private foundation interested in our forward movement. We then identified about one hundred members of the Garden staff—about one-third of our employees—to participate in the process. We called them the Expanded Planning Group. These individuals were drawn from senior management (the people reporting to me); middle management (those reporting to the senior managers); and frontline management (the people out in the Garden greeting the public, scientists applying to the National Science Foundation for grants, instructors in the education department, and weekend security supervisors). I have always felt that these frontline managers (my favorite example is the woman who runs the Everett Children's Adventure Garden on weekends) know more about what's happening and what would be valuable new programming than many of us in administrative offices. We identified these strata of management as Level I (Frontline Managers); Level II (Middle Managers); Level III (Senior Managers); and Level IV (the CEO, that's me, and the Board Committee for Strategic Planning).

Having drawn what amounted to a special table of organization for the purposes of planning, we then set aside dates for two-hour sessions of presentations of a few minutes each by Level I, frontline managers, to the entire Expanded Planning Group. Each of these people (I would guess there were forty to forty-five at this level) had been asked to write down their thoughts about programming in their area over the seven-year period identified. We devised forms they could fill out, and these forms were divided into three categories: "Business as Usual" over seven years; "Better Business and Recommended Improvements"; and "Blue Sky Thinking"—what would be marvelous? We made it clear to these frontline managers that they were to develop their thinking in discussion and consultation with the middle managers to whom they reported. They were not to invent their own material without vetting it with the person right above them. The facilitator worked with every one of these individuals to help them clarify their thinking, fill out their forms, and create short oral presentations with a few slides.

Seasoned institutional and corporate managers will understand from this description that the people being asked to make these presentations were not experienced at public speaking, and, of course, they were nervous about it. But the facilitator coached them through rehearsals, and the exercise was a fabulous learning experience for the lower-level managers. Many of these people—for example, the woman running the Children's Adventure Garden on Sunday afternoon—came up with extremely valuable and interesting ideas because they all knew their work so well. So, over several weeks, these two-hour meetings took place, perhaps one per week, and we would review the program and the programmatic needs of the entire organization, wall to wall, through the eyes and experience of these frontline managers.

PhD scientists would sit through presentations by instructors dealing with elementary school children. Fundraising staff would sit through presentations delivered by security and operations personnel. Everyone at the institution began to understand and appreciate what their colleagues in other divisions were doing. The board members who attended these sessions (the twelve members of the Strategic Planning Committee and all other board members were invited) were ecstatic. They loved hearing an assistant manager of public relations or a head of weekend maintenance talk about their work and problems.

In the weeks following the Level I presentations, the middle managers at Level II were asked to make theirs, and it was at this level that the staff began to set priorities. After all, the director of children's education, who was a middle management person, could not simply repeat all we had heard from her direct reports. She had to begin to set her priorities. And she was asked to do that in consultation with the vice president to whom she reported. It was also at this level that I began to be involved in the thinking, because the vice presidents would come to me with summaries of what they thought their middle managers were going to present. There were very few surprises, and therefore very few arguments or hitches in the proceedings.

After we had reviewed the whole organization at Level II, Level III (senior management) presentations followed, and these began to be highly strategic with priorities set because there was clearly too much on the table. We recognized that it could never all be done. Once again, those of us present at these meetings (remember, the entire Extended Planning Group of one hundred staff was present for every one of these meetings) heard programmatic description and wishful thinking across the whole organization. So, by the time the Level III presentations were concluded, we had all been exposed to programmatic thinking across the board, *three times.* This iterative process is a way to make decisions and achieve considerable consensus among all the participants.

My Level IV presentation as CEO set top priorities and was the exciting conclusion of this process. The CEO has the serious responsibility to get it right, make it interesting and not too detailed, synthesize and summarize it in ways that are not boring but, rather, energizing. The CEO has to sell their version of these ideas as they have filtered upward through everyone. This first "sell" is to staff and board, and now a picture emerges of where the CEO and staff and trustees want the organization to be in year eight, when the plan has been accomplished.

As the exercise is unfolding, vice presidents and the finance office have to create the matching plan for capital improvements required by the programmatic recommendations. And the finance office has to cost out everything in a comprehensive seven-year financial plan. And the fundraisers have to think about how to raise more money than they have in the past. And the board has to be excited enough to back the plan enthusiastically. The whole team has to be mobilized, and this is again the responsibility of the CEO.

It's instructive to think how much the results of such a planning process can do for a single area in an organization. For example, in plant research and conservation our detailed planning enabled us to build about ninety million dollars' worth of new scientific facilities, funded by enthusiastic individual donors, the foundation community, and the federal government; to create seventeen new endowed chairs for scientists, funded entirely by passionate individual donors; to invent molecular and genomics research programs, which have received over $45 million from the federal government and private donors; to digitize and image millions of herbarium specimens and books for universal access on the internet, at a cost of tens of millions and funded by major private foundations and corporations such as Google. Not to mention, to support $10 million in annual operating expenses covering scientific salaries, expeditions, and programs such as graduate studies. All of this was born in the planning process, and the funders were people and agencies who learned to love and respect science as a result.

This planning design was repeated four times in our generation at the Garden. The work carried out in each of the four periods that followed the publication of the plans is the story of this book.

Acknowledgments

Just as our work at NYBG was teamwork all the way, so also has been the writing of this book. For their devotion to NYBG, there are many individuals to be recognized. First, those who were my main partners on the board over so many years. In chronological order, my chairs Jim Ferguson, Tom Hubbard, Roly Nolen, and Maureen Chilton all cared deeply about what we were doing and backed me completely. And at that level, of course, the inimitable Enid A. Haupt, who saved the Garden with her loving bequest.

In addition to the marvelous trustees I inherited and to later members of the board connected with major projects in my twenty-nine years, who are mentioned throughout the text, I particularly want to thank current trustees Leonard Abess, Ed Bass, Fredericke Biggs, Susan Burke, Whitney Clay, Barclay Collins, Florence Davis, Edith Everett, Lionel Goldfrank, Diane Katzin, George Milne, Diane Renwick, Barbara Robinson, Janet Ross, and Julie Sakellariadis, along with the late Honoré Wamsler, for their long and creative service in multiple roles over these decades.

Special thanks also to the family of Robert Zoellner for their loyal and generous support of the *Holiday Train Show* since the beginning; to Larry Lederman the attorney for his wise legal counsel and to Larry Lederman the photographer for the two beautiful books he created for us; to Bunny Williams and John Rosselli for giving us pro bono the murals and the elegant decorating in the Garden Terrace Room; and to Kate Levin, former New York City Commissioner of Cultural Affairs, for her interest in NYBG over many years.

Often mentioned in the foregoing text but deserving to be called out again are my stellar compatriots John Rorer, J.V. Cossaboom, and Todd Forrest. Fifteen years after Enid Haupt, LuEsther Mertz and her trustees—Larry Condon, Georgia Delano, John Hoffee, Bill O'Connor, Paul Richards, and Robert Williams—saved the Garden again with their enlightened philanthropy.

In charge of all the massive fundraising we did over this time were four development officers who particularly distinguished themselves, John Bernstein, Rosemarie Garipoli, Tim Landi, and Jennifer Rominiecki. We have always been much in their debt.

Now for help with the book. Thanks very much indeed to our friend in London Robin Lane Fox, the distinguished classicist and garden writer (forty years of weekly columns in the *Financial Times*), for his enthusiastic foreword. He knows the Garden well and loves the role of botanical gardens in the world. No one could have been more appropriate—or generous.

The wonderful Leon Levy Foundation has supported the research and typing, and also helped with the costs of publication. So have John Bernstein, Susan Burke, Larry Condon, and two anonymous donors—you know who you are.

I have tried not to bother staffers at NYBG, because this is not an official history, and I know how much they have had on their minds and plates during the pandemic, but I wish to thank four current or former staff members who have read all or part of the text and made essential comments and corrections and whose fact-checking was invaluable: Brian Boom, Susan Fraser, Stephen Sinon, and Barbara Thiers. Trustees Ken Roman and Caroline Wamsler also reviewed early drafts and were insightful. The illustrations have been researched and organized by Marlon Co, Christopher Kozarich, and Lauren Turchio at the Garden with input from Todd Forrest, and Marlon took many of the beautiful NYBG photographs. For much cheerful support of many kinds over the years, thanks are due to Barbara Corcoran, Melinda Manning, and Bergita Nokaj.

In chapter 4, I quote the phrase "Rainforest Remedies." With their permission I have borrowed this colorful title from an excellent book about ethnobotanical research in Belize by Michael Balick and Rosita Arvigo. I have always depended on the knowledge and expertise of my colleagues, and that's always been part of the fun.

Many friends outside NYBG have guided my thinking as I debated about how to write this all down. They include Neil Baldwin, Peg Breen, Jane Garmey, Roberta Gratz, Marilan Lund, Elizabeth Moynihan, Nancy Nicholas, Phyllis Rose, Elizabeth White, Carey Winfrey, and my mother, Elizabeth Dunn.

The text has been researched in part and typed over and over and over again by Sara DuPont. She is a most excellent collaborator. Sara used to work at NYBG, but she now lives with Jules and baby Charlie in Austin, Texas. During our long telephonic and electronic relationship, we have never seen each other!

Robin Karson, executive director of the Library of American Landscape History, LALH senior editor Sarah Allaback, managing editor Carol Betsch, and associate director and designer Jonathan Lippincott have all labored mightily to make this book "as good as it can possibly be," to quote Robin. Thanks to them for making it as good as it can possibly be, and beautiful, and for putting up with me.

The co-captain of this team is of course Scott Newman. He has been a wise and patient sounding board and given me sage advice about many details better left out! He was part of the NYBG family through almost all of the years described in these pages and was always an enthusiastic participant. He has been the most creative and best adviser imaginable. How many tricky speeches have I rehearsed by reading aloud to him over a glass of wine in our West Village kitchen? He is also VP for IT in this household. What a gift his love has been.

I have now finished telling these stories. That was a fateful morning in Copenhagen.

Further Reading

125th Anniversary Strategic Plan. Bronx: New York Botanical Garden, 2016.

All Aboard! A Tour of the Holiday Train Show at the New York Botanical Garden. Bronx: New York Botanical Garden, 2010.

Anderson, Frank J. *A Treasury of Flowers: Rare Illustrations from the Collections of the New York Botanical Garden*. Boston: Little, Brown, 1990.

Before the Green Is Gone: A New Century of Biodiversity Research at the New York Botanical Garden. Bronx: New York Botanical Garden, 2005.

Boyer, James. "Scientific Adventure in a Children's Garden." *Public Garden: The Journal of the American Public Gardens Association* 28, no. 3 (2014): 44–56.

Britton, Nathaniel Lord. *History of the New York Botanical Garden*. Bronx: The Garden, 1915.

Callery, Bernadette G. "Collecting Collections: Building the Library of the New York Botanical Garden." *Brittonia* 47, no. 1 (1995): 44–56.

Daly, Douglas C., and Elizabeth A. Christy. *Neighborhood Composting in New York City*. New York: Council on the Environment of New York City, c. 1978.

Eustis, Elizabeth S. *European Pleasure Gardens: Rare Books and Prints of Historic Landscape Design from the Elizabeth K. Reilley Collection*. Bronx: New York Botanical Garden, 2003.

A Forest in the City: Centuries of Sylvan Beauty. Bronx: New York Botanical Garden, 2011.

Forrest, Todd, and Jessica Arcate Schuler. *Thain Family Forest Management Plan*. Bronx: New York Botanical Garden, 2015.

Fox, Robin Lane. "The Floral Kingdom in the Bronx." *New York Review of Books*, January 8, 2015, 652–54.

Fraser, Susan M., and Vanessa Bezemer Sellers. *Flora Illustrata: Great Works from the LuEsther T. Mertz Library of the New York Botanical Garden*. New York: New York Botanical Garden; New Haven: Yale University Press, 2014.

Glenn, William H. *Guide to Geology: The New York Botanical Garden*. Bronx: New York Botanical Garden, 1978.

Grimes, William. "Kahlo's Primal Eden: Replanted in the Bronx." *New York Times*, May 22, 2015.

Heritage Landscapes, *The New York Botanical Garden Cultural Landscape Report: Landscape History*. Charlotte, VT: Heritage Landscapes, 2008.

Holmgren, Patricia, Jacquelyn Kallunki, and Barbara Thiers. "A Short Description of the Collections of the New York Botanical Garden Herbarium (NY)." *Brittonia* 48, no. 3 (1996): 285–96.

Hubbard, Juliet Alsop. *A Walk in the Wild: Native Plants and Their Habitats at the New York Botanical Garden*. Bronx: New York Botanical Garden, 1991.

Johnson, Ken. "Giverny Blooms in the Bronx." *New York Times,* May 18, 2012.
Johnson, Victoria. *American Eden: David Hosack, Botany, and Medicine in the Garden of the Early Republic.* New York: Liveright, 2018.
Kimmelman, Michael. "Where the Art Grows on Trees (and Everywhere Else)." *New York Times,* December 30, 2005.
Klinkenborg, Verlyn. "Visit the New York Botanical Garden Now, and Learn about the Potent Appeal of Bare Nature." *Town & Country,* December 2005, 235–42.
Kohn, David. *Darwin's Garden: An Evolutionary Adventure.* Bronx: New York Botanical Garden, 2009.
Lacayo, Richard. "Top Ten Museum Exhibits: Moore in America." *Time,* November 3, 2008.
Lederman, Larry, and Todd A. Forrest. *Magnificent Trees of The New York Botanical Garden.* New York: Monacelli Press, 2012.
Liedtka, Jeanne M. *Strategic Planning at the New York Botanical Garden (A).* Darden School of Business, February 11, 1998.
____. *Strategic Planning at the New York Botanical Garden (B).* Darden School of Business, February 11, 1998.
____, and Randy Salzman. *Strategic Planning at the New York Botanical Garden (C).* Darden School of Business, January 20, 2008.
Long, Gregory. "The New York Botanical Garden." *Site/Lines* 2, no. 1 (Fall 2006): 13–15.
____, and Todd A. Forrest. *The New York Botanical Garden.* New York: Abrams, 2016.
Mickulus, Peter Philip. *Britton's Botanical Empire: The New York Botanical Garden and American Botany, 1888–1929.* Bronx: New York Botanical Garden Press, 2007.
Mullaly, John. *The New Parks beyond the Harlem: Nearly 4,000 Acres of Free Playground for the People.* New York: Record & Guide, 1887.
A New Strategic Plan: Into the 21st Century 2009–2015. Bronx: New York Botanical Garden, 2008.
The New York Botanical Garden: The Financial Plan, 1993–1999: Financial Forecast for the New York City Campus, prepared by John Rorer. Bronx: The Garden, 1992.
The New York Botanical Garden: The Fundraising Plan, 1993–1999: Achieving the Program Priorities of the New York City Campus, prepared by Rosemarie Garipoli. Bronx: The Garden, 1992.
The New York Botanical Garden: The Master Plan, Phase I, 1993–1999, prepared by R. Geoffrey Roesch. Bronx: The Garden, 1992.
The New York Botanical Garden: Plan for a New Era, 2001–2007. Bronx: The Garden, 2000.
OLIN. *Master Plan. Site Inventory and Analysis.* Philadelphia: Olin Partnership, 2008.
Olmsted Brothers. *Report on the New York Botanical Garden,* 1924.
Rothstein, Edward. "Grandes Dames of the Gardens." *New York Times,* May 16, 2014.
Ruggiero, Michael, and Ruth Rogers Clausen. *Perennial Gardening.* New York: Pantheon Books, 1994.
Science at the New York Botanical Garden. Bronx: New York Botanical Garden, 2008.
Thiers, Barbara M. *Herbarium: The Quest to Preserve and Classify the World's Plants.* Portland, OR: Timber Press, 2020.
Tomasi, Lucia Tongiorgi. *The Renaissance Herbal.* Bronx: New York Botanical Garden, 2013.
Utterback, Christine, and Michael Ruggiero. *Reliable Roses.* New York: Clarkson Potter, 1997.
Zavala, Adriana, Mia D'Avanza, and Joanna L. Groarke, eds. *Frida Kahlo's Garden.* New York: New York Botanical Garden and DelMonico Books / Prestel, 2015.

Index